Jan de Witt's Elementa Curvarum Linearum
Liber Secundus

For other titles published in this series, go to
http://www.springer.com/series/4142

Sources and Studies
in the History of Mathematics and
Physical Sciences

Jan de Witt's Elementa Curvarum Linearum

Liber Secundus

Edited by

Albert W. Grootendorst
Jan Aarts
Miente Bakker
Reinie Erné

Editors
Albert W. Grootendorst †
Formerly Professor of Mathematics
Delft University of Technology
Delft
The Netherlands

Jan Aarts
Formerly Professor of Mathematics
Delft University of Technology
Delft
The Netherlands
johannesaarts@gmail.com

Miente Bakker
Formerly Researcher
and Scientific Administrator
Centrum Wiskunde & Informatica
Amsterdam
The Netherlands
mientedbakker@gmail.com

Reinie Erné
Leiden
The Netherlands
contact@ernemath.com

ISBN 978-1-4471-2597-6   ISBN 978-0-85729-142-4 (eBook)
DOI 10.1007/978-0-85729-142-4
Springer London Dordrecht Heidelberg New York

British Library Cataloguing in Publication Data
A catalogue record for this book is available from the British Library

Mathematics Subject Classification (2010) 01A45 (01A75 51-03)

Printed on acid-free paper

Springer is part of Springer Science+Business Media (www.springer.com)

piae memoriae
uxoris meae

Jan de Witt (1625–1672)
Portrait by **Jan de Baen** (20 February 1633, Haarlem – 8 March 1702, Den Haag)
Rijksmuseum Amsterdam

Preface by the editors

The first textbook on analytic geometry was written in Latin by the Dutch statesman and mathematician Jan de Witt. It is entitled *Elementa Curvarum Linearum* [*Foundations of curved lines*] and consists of two volumes. The first volume, *liber primus*, presents the geometric generation of the conics. The second volume, *liber secundus*, deals with the classifications of quadratic curves. The second volume forms the core of the *Elementa Curvarum Linearum*, while the first volume only serves as an introduction. The basis for this textbook, which was first published in 1659, lies in the ideas that Descartes developed in his book *Géométrie* (1637).

Albert Grootendorst prepared the English translation of the first volume, *liber primus*, which was published by Springer in 2000. In addition to the parallel presentation of the Latin text and its English translation, this includes a general introduction, a summary with a complete survey of the theorems (without proofs), annotations, and two appendices. This edition is partly based on the Dutch translation by Grootendorst, which was published in 1997 by the Centrum Wiskunde & Informatica in Amsterdam. The Dutch translation of the second volume, *liber secundus*, by Grootendorst, was published in 2003 by the Centrum Wiskunde & Informatica. This book has the same format as both the Dutch and English translation of the first volume: in addition to the text and its translation, it includes a general introduction, a summary of the theorems, annotations, and an appendix.

Grootendorst received much help from Miente Bakker in preparing the publication of both the Dutch and English translations. Owing to his premature death in December 2004, Grootendorst was unable to complete the English edition of the second volume. At the time of his death, Grootendorst had almost completed the translation of the Latin text into English. His plan was also to translate the additional material of the Dutch edition into English. Jan Aarts completed the translation of the Latin text into English. Reinie Erné translated the additional material from the Dutch edition into English. Miente Bakker did the editorial work, including preparing the index.

The Latin text that is used for this edition of the *liber secundus* is taken from the second edition of 1863 by the publisher Blaeu in Amsterdam; we are indebted to K.F. van Eijk, treasurer of the library of the Delft University of Technology, for providing access to it. Many thanks go to Tobias Baanders of the Centrum Wiskunde & Informatica for the additional figures illustrating the introduction, summary and annotations.

In concluding, we give the last paragraph of Grootendorst's Preface of the Dutch edition of the *liber secundus*, adapted to the present situation:

> This marks the completion of the English translation of the *Elementa Curvarum Linearum*, the magnum opus of the Dutch statesman Jan de Witt who was one of the greatest mathematicians of the 17th century and who might

have been the greatest, had he not been distracted by so many state affairs. To quote Christiaan Huygens:

Nullam aeque saeculum geometrarum ferax fuisse arbitro, inter quos vir ille, si negotiis minus distringeretur vel principem locum obtinere posset. [In my view no century has been so rich in mathematicians, amongst whom this man (J. de Witt) might have taken the first place, had he been less distracted by state affairs.]

We hope that this translation makes the work of the great scientist Jan de Witt accessible to a broad group of today's mathematicians.

Amsterdam, June 2010

Jan Aarts, Miente Bakker and Reinie Erné

Contents

François Viète
(1540–1603)

Frans van Schooten, jr.
(1615–1660)

Pierre de Fermat
(1601–1665)

René Descartes
(1596–1650)

Albert Grootendorst (1924–2004) studied mathematics at Leiden University, where he obtained his PhD for a dissertation on Theta-series in relative-quadratic numbers (1958); his supervisor was Professor H.D. Kloosterman. From 1952 to 1956 he was a teacher at a grammar school in the Hague. From 1956 to 1989 he was attached to Delft University of Technology as a member of the scientific staff, a lecturer, and then a full professor. In cooperation with his colleague Professor B. Meulenbeld he wrote a textbook on analysis in three volumes. He also translated a number of seventeenth-century mathematical texts from Latin into Dutch and into English, and wrote a number of articles on the history of mathematics. With Professor W.C. Waterhouse and Dr C. Greither he revised Clarke's translation of *Disquisitiones Arithmeticae* by C.F. Gauß. In 1997 and 2003 he published annotated Dutch translations of *Elementa Curvarum Linearum – Liber Primus* and *Elementa Curvarum Linearum – Liber Secundus* by Jan de Witt. In 2000 he published the annotated English translation of *Liber Primus*.

1

Introduction

1.1 The second part of Jan de Witt's *Elementa Curvarum Linearum* is the essence of the whole work. The first part was merely a necessary preparation to *Liber Secundus*, which is referred to in the correspondence and in Part I as the *tractatus* (or *compositio*) *locorum planorum et solidorum*.

This last text was delivered to Van Schooten at the beginning of 1658. In a letter to Jan de Witt dated 8 February 1658, Van Schooten expressed his great appreciation for this work and promised to study it carefully and to help in any way he could with the preparation for publication.

On 6 October 1658, Jan de Witt received the results of Van Schooten's efforts, with an accompanying letter in which he wrote that

> ... *so veel 't mij doenlijck geweest is, accuraet (heeft) naergesien...* [... I have checked (it) as carefully as I could...]

In particular, the following passage is important:

> *Hebbe in het uytschryven de voorszeide calculatie op monsieur Des Cartes manier gestelt en op eenige weynige plaetsen de woorden wat verandert, om doorgaens, so veel 't mogelijck was, overal de tael sijnder geometrie, die nu by meest alle de fraeyste verstanden de allerbekendste is, te gebruycken...* [In my exposition, I have written the aforementioned computation in the manner of Mr. Des Cartes and have changed the words in a few places in order to use, as much as possible, the language of his geometry, which is now the most current among almost all distinguished minds...]

Jan de Witt immediately replied with a letter of 8 October, in which he expressed his gratitude, but added that he had no time to spend on this work. Nevertheless, he emphasized that this text

> *Niet anders en mach voor de dach comen dan voorhenen gaende eene corte verhandelinge van de nature ende proprieteyten der cromme liniën.* [Must not be published without being preceded by a brief treatise on the nature and properties of curved lines.]

Apparently he had already finished this brief treatise, because he joined it to the letter with the request to

> ... *insgelijkx eens te doorsien ende de faulten ... te verbeteren* [... look it over and correct any mistakes...]

A.W. Grootendorst et al. (eds.), *Jan de Witt's Elementa Curvarum Linearum*,
Sources and Studies in the History of Mathematics and Physical Sciences,
DOI 10.1007/978-0-85729-142-4_1, © Springer-Verlag London Limited 2010

In the final publication this treatise became *Liber Primus*; it gives a mechanical description of the known conics as plane curves, *absque ulla solida consideratione*: that is, without any spatial considerations.

The style of *Liber Primus* is clearly different from that of *Liber Secundus*. Indeed, in the first part the method of Descartes is not used at all: all calculations are done according to the geometric algebra of Euclid (as explained in Note [3.30]; see also [26]).

The definitions given there serve as the basis for *Liber Secundus*, whose core can be described as the characterization of the conics by means of equations in two variables x and y which can be seen as the coordinates (which are line segments) of the points on the curves, and the deduction of the conics' properties from this, all using the analytic method.

This results in a tightly ordered enumeration of *Theoremata* (theorems) and *Problemata* (problems). We will come back to the structure of this work in Section 1.10 of this introduction. It is written more in the rigid style of the *Elements* of Euclid than in the style of the *Géométrie* of Descartes and can rightly be considered the first systematic textbook of analytic geometry in the Cartesian tradition. The first book that can be seen as a successor of this work is *Elementa Matheseos Universae* of Christian von Wolff (1679–1754).

To help understand the significance of *Liber Secundus*, we will attempt to place it in the context of its time by giving some well-known, relevant facts.

1.2 Whenever the origin of analytic geometry is brought up, the names of René Descartes (1596–1650) and Pierre de Fermat (1601–1665) immediately come to mind. Their sources of inspiration, however, lie in a more distant past: the Greek antiquity, with names such as Euclid (ca 300 BC), Archimedes (287–212 BC), Apollonius of Perga (second half of the third century BC), and Pappus of Alexandria (first half of the fourth century AD). Moreover, the tools that were implicitly passed on to them, and which helped make their results possible, date mostly from the late Middle Ages and the Renaissance. Of that time only Nicole Oresme (ca 1320–1382) and François Viète (1540–1603) are mentioned here.

1.3 The *Collectio* (Συναγωγή) of Pappus occupies a central place in the developmental history of analytic geometry. Several causes can be given for this.

This collected work consisting of eight books and giving an extensive overview of the work of some thirty mathematicians, from Euclid to Pappus's contemporary Hierius, was composed by a capable mathematician. It derives its importance not only from the discussion of works known to us, but also in particular from the sometimes concise remarks concerning writings that are now lost. This is the most important aspect of the *Collectio*; because of this it led to and provided support for the reconstruction of these lost works.

Because of the renewed interest in Greek and Roman culture, including mathematics, which flourished in the Renaissance, much attention was given to this reconstruction in the sixteenth and seventeenth centuries.

Important contributions were made by the Frenchman Viète with his *Apollonius Gallus* (1600), by the Dutchman Snellius with his *Apollonius Batavus* (1607/1608), by the Italian (by origin) Ghetaldi with his *Apollonius Redivivus* (1607/1613), as well as by Fermat with his reconstruction of the *Loci Plani* of Apollonius, which he presented to his friend Prade around 1630 and which was one of his sources of inspiration in his setting up of analytic geometry. Frans van Schooten Jr. also attempted to reconstruct the *Loci* of Apollonius. We find the results of this attempt in his *Excercitationum Mathematicarum Libri V* (1656/1657).

In this respect, the seventh book of the *Collectio* deserves particular attention. It not only contains many theorems (more than 400) from lost work, but also a definition of the fundamental terms analysis (αναλυσις) and synthesis (συνθεσις). It is known as the treasury of analysis (τοπος αναλυομενος).

In the opening words of this work, dedicated to his son Hermodorus, Pappus describes analysis as a subject that is intended for those who already master the well-known *Elements* and now wish to study problem solving.

Pappus names Euclid, Apollonius, and Aristaeus the Elder (ca 350 BC) as founders of the analytic method. Heath, however, suspects that the term analysis was already known in the school of Pythagoras (see [33]).

Diogenes Laertius (third century AD) attributed the term to Plato, but it is probable that Plato laid the emphasis rather on the associated synthesis. In the synthetic setting up of the *Elements* of Euclid the term analysis does not come up explicitly, but it may have played a role in the heuristics. In some manuscripts of the *Elements* it appears in a marginal comment; Heiberg assumes that this is an insertion by Hero (ca 60 AD), which refers to the research of Theaitetus (410–368 BC) or of Eudoxus (408–355 BC).

Right after the introduction Pappus gives a definition of analysis and of synthesis, as follows:

> Analysis is a method where one assumes that which is sought, and from this, through a series of implications, arrives at something which is agreed upon on the basis of synthesis; because in analysis, one assumes that which is sought to be known, proved, or constructed, and examines what this is a consequence of and from what this latter follows, so that by backtracking we end up with something that is already known or is part of the starting points of the theory; we call such a method analysis; it is, in a sense, a solution in reversed direction. In synthesis we work in the opposite direction: we assume the last result of the analysis to be true. Then we put the causes from analysis in their natural order, as consequences, and by putting these together we obtain the proof or the construction of that which is sought. We call this synthesis.

Let us add two remarks. First of all a linguistic observation: the word analysis is described by Pappus as *anapalin lusis (ανα παλιν λυσις)*, which is the Greek term for solution in reversed order; synthesis (συνθεσις) means composition.

Next let us point out the remarkable manner in which analysis is described. If we indicate the stages of the analysis with $a_1, a_2, \ldots$, where a_1 is that which is sought, then, formally, Pappus's reasoning is not $a_1 \rightarrow a_2$ but $a_1 \leftarrow a_2$, where a_2 is a sufficient condition for a_1.

Pappus distinguishes two types of analysis:

1.   *Theoretic* or *zetetic* analysis (from *zèteo*, to search for), which concerns the truth of a statement.

2.   *Problematic* or *poristic* analysis (from *porizo*, to provide; a porism is halfway between a problem and a theorem), which concerns the constructability or computability.

When giving the definition of these two types of analysis, Pappus further remarks that for these we must assume that which is sought to be true or possible, and then, by drawing logical conclusions, end up with something known to be true or possible or known to be untrue or impossible. In the first case we must then follow the reasoning in reverse order to conclude that the assumption was correct. With this, Pappus stresses that each step in the reasoning must be reversible. In the second case the assumption is, of course, incorrect.

Viète would give other meanings to these terms and distinguish a third type of analysis: the *rhetic* or *exegetic* analysis. We will come back to this later.

Yet, the *Collectio* not only derives its significance for the development of analytic geometry from the references it contains and from the attention it gives to the terms analysis and synthesis, but also in particular from problem it states, which would become known as one of the Problems of Pappus.

With modern notation we can state this problem as follows:

Given three or four straight lines l_1, l_2, and $l_3 (l_4)$ in the plane, we ask for the set of points P whose distances from d_1, d_2, and $d_3 (d_4)$ satisfy

(i) $d_1 d_2 : d_3^2$ = constant   (in the case of three lines),

(ii) $d_1 d_2 : d_3 d_4$ = constant   (in the case of four lines).

These distances can be taken perpendicularly, but also in a direction that is determined separately for each line l_i. It is clear that this choice does not change the nature of the problem.

A possible generalization to more than four lines is obvious and is already mentioned by Pappus himself (*Collectio* vii, 38–40; ed. Hultsch [36], p. 680, and I. Thomas [57], Part II, pp. 601–603).

In the case of five lines $l_1, l_2, ..., l_5$, the distances are $d_1, d_2, ..., d_5$.

Pappus considers the two parallelepipeds that are enclosed by d_1, d_2, d_3 and by d_4, d_5 and a randomly chosen line segment a. The condition on P is now that the ratio $d_1 d_2 d_3 : a d_4 d_5$ is constant.

For six lines the distances are $d_1, d_2, ..., d_6$. The condition is then that the ratio $d_1 d_2 d_3 : d_4 d_5 d_6$ is constant. The case of four lines is shown in Figure 1.1. Let us already mention that in this case the loci in question will prove to be conics.

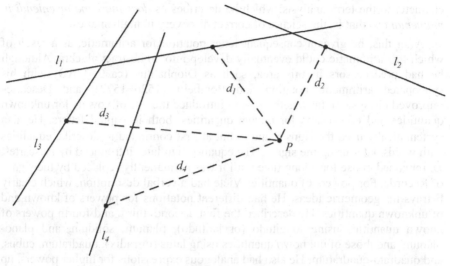

FIGURE 1.1

In the general case of an even number of lines, say $2n$, the condition becomes that the ratio

$$d_1 d_2 ... d_n : d_{n+1} d_{n+2} ... d_{2n}$$

is constant and in the case of an odd number of lines, say $2n+1$, the condition becomes that the ratio

$$d_1 d_2 ... d_{n+1} : a d_{n+2} d_{n+3} ... d_{2n+1}$$

is constant, where a is again a randomly chosen line segment.

For the context of this problem we refer the reader to the appendix, Section 5.1.

1.4 Of those who were inspired by the *Collectio* of Pappus, in particular Book VII, we first mention François Viète (1540–1603), who worked as a lawyer for the government and in his spare time was an avid mathematician.

We mentioned him above in relation to his *Apollonius Gallus*, a reconstruction of works of Apollonius, and also in relation to his views on the terms analysis and synthesis as presented by Pappus.

His most important mathematical work is considered to be *In artem analyticem isagoge*, which was published in 1591 in Tours. This work begins with a chapter on the term analysis, which already came up for discussion above. Instead of the bipartition of Pappus, he suggests his own division into three parts. For him *zetetic* analysis is the determination of an equation or proportion (with known coefficients) that the unknown quantity must satisfy. In subsequent computations this quantity is considered known. In *poristic* analysis, the correctness of a theorem is studied using this equation or proportion, while in *rhetic* or *exegetic* analysis the unknown quantity is deduced from it. In fact, Viète gives an algebraic character to the term analysis, which he describes as *doctrina bene inveniendi in mathematicis*: that is, the science of correct discovery in mathematics.

After this, he gives a consequent letter notation for arithmetic, as a result of which this arithmetic could eventually develop into an abstract algebra. Although he had predecessors in this area, such as Diophantus (ca 250 AD) with his syncopated arithmetic notation and Bombelli (1526–1572), and Descartes improved his system, he was the first to introduce the use of vowels for unknown quantities and consonants for known quantities, both in capital letters. He also systematically used the signs + and −, but at first continued to describe equalities with words, later using the sign ∼. The equality sign later introduced by Descartes, ∝, remained in use for a long time until it was permanently replaced by the sign = of Recorde. For powers of quantities Viète had a verbal description, which clearly betrays his geometric ideas. He had different notations for powers of known and of unknown quantities. He described the first, second, third, and fourth powers of known quantities using longitudo (or latitudo), planum, solidum, and plano-planum, and those of unknown quantities using latus (or radix), quadratum, cubus, and quadrato-quadratum. He also had analogous expressions for higher powers, up to and including the ninth power.

An example:

$$B\,A \text{ quadratum} + C \text{ planum } A \text{ aequalia } D \text{ solido,}$$

corresponds to our:

$$bx^2 + c^2 x = d^3.$$

His geometric interpretation of products not only is reflected in his notation, but also forces him to use homogeneous formulas because, according to a condition of Aristotle, only similar quantities can and may be compared to each other. The terms of an equation are also called the *homogenea*.

It is remarkable that he speaks of powers higher than the third power in spite of the fact that these have no geometric meaning. Descartes would solve this dilemma later. Of course the innovations of Viète include more than those we mention here. These, however, are of exceptional importance: this is where the distinction arises between *logistica numerosa*, computing with explicit numbers, and *logistica speciosa*, computing with letters. As a result one could speak of equations in general terms and no longer had to rely on specific examples. Moreover, thanks to the *logistica speciosa*, the dependence of the solution on the coefficients of the equation becomes clear.

Viète used his notations in particular to solve geometric problems. To this end he transformed the problem into an algebraic equation in one variable, which he solved using his new technique, whereupon he constructed the solution, whenever possible. The "construction of equations" (that is, their geometric resolution) was henceforth done by means of algebra instead of the geometric and verbal techniques of antiquity (see also [26] and [27]). A well-known and much cited example is the determination of the sides of a rectangle when their ratio and the area of the rectangle are given.

Of course a construction with ruler and compass only succeeded if the equation was of degree at most two. Equations of degree three or four could be solved algebraically, but their roots could in general not be constructed with these instruments. For these cases and others Viète suggested the use of other tools. In this he did not follow the method already used by Menaechmus (ca 350 BC), namely using common points of curves. As an example of that we give, in our notation, the method used by Menaechmus to find both geometric means x and y of a and b.

From $a : x = x : y = y : b$ follow $x^2 = ay$ and $y^2 = bx$ (and also $xy = ab$), so that $x^4 = a^2 y^2 = a^2 bx$ and therefore $x^3 = a^2 b$. Menaechmus solved this equation by intersecting two conics he had discovered, of which he knew the geometric properties. In modern terms: he intersected the parabola determined by $x^2 = ay$ with the parabola characterized by $y^2 = bx$. Let us already mention at this point that Descartes and Fermat would continue using this method with curves of higher degree.

The above clearly shows that Viète did not get around to analytic geometry: he did not draw curves other than a straight line or a circle, and did not use coordinate systems. But the most important shortcoming is that he restricted himself to so-called "determinate" equations, that is, equations in one variable with constant coefficients. He did not know equations in two variables, as a result of which he could not describe *loci* (τοποι) algebraically, where a *locus* is the set of all points whose location is determined by stated conditions.

1.5 The first fundamental contribution of Descartes to the foundations of analytic geometry is the creation of its own algebraic apparatus. This includes a notation that we still use today, except for the equal sign for which, as mentioned before, Descartes chose ∞. In this, though not only in this, he surpassed Viète, a fact of which he was aware: *...je commence en cela par ou Viète a finy.* [...in this matter, I begin where Viète has left off].

In short, the innovation of Descartes consists of defining addition, subtraction, multiplication, and division, hence also power taking and root extraction, for line segments in such a way that their results are once more line segments, so that the set of line segments is closed under these operations. As we already saw in *Liber Primus*, this was not the case for the mathematics of the Greeks: the sum and difference of line segments were indeed line segments, but the product of two line segments was a rectangle and the product of three was a rectangular parallelepiped. The product of more than three line segments was problematic (for this, see Appendix, Section 1).

Descartes presents his ideas on the first page of the *Géométrie*. It is essential for this that he introduces a fixed line segment that will serve as the unit element 1. The proportions $1 : a = b : p$ and $1 : b = q : a$ give the product $p = ab$ and the quotient $q = a/b$. Figure 1.2 gives the associated constructions.

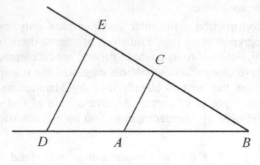

FIGURE 1.2

Here we have $AB = 1$ and $AC \parallel DE$, so that $1 : BD = BC : BE$, and therefore

$$BE = BD \cdot BC \quad \text{and} \quad BC = BE/BD .$$

For the extraction of roots Descartes uses a well-known property of the similar triangles formed by the altitude from the right angle in a right triangle, and gives Figure 1.3 as an example, where triangle *IFG* is similar to *HIG* and consequently $IG^2 = FG.GH$.

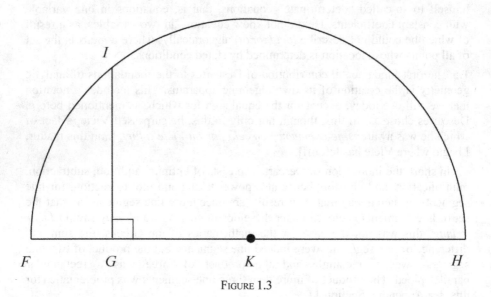

FIGURE 1.3

An important consequence of the introduction of the unit line segment is that Descartes can drop the condition of homogeneity. After all, all products and quotients of line segments are also line segments. Viète still required that all quantities in an equation or equality be of the same type: all line segments, all areas, or all volumes. For Descartes, however, $a^2b^2 - b$ is by definition a line segment, though he does add that, for example when extracting the cube root, this form can be written as a polynomial of degree three, that is, as

$$\frac{a^2b^2}{1} - 1^2b.$$

It is remarkable that Descartes nevertheless still speaks of the product of three line segments as *le parallelepipede composé de trois lignes* [*the parallelepiped made up of three lines*], for example in the *Géométrie* p. 336.

Right after this Descartes introduces the following notation, which is so familiar to us today: *a, b, c* ... for known line segments and ... *x, y, z* for unknown line segments. Moreover, he also introduces numerical exponents and the radical sign. We have already mentioned his variant of the equal sign. Van Schooten refers to these innovations as *the manner of Monsieur Des Cartes*.

As a first application, Descartes gives the constructive resolution of the quadratic equation $z^2 = az + b^2$. As usual he restricts himself to the positive root, dismissing the negative one as *racine fausse* (false root).

Figure 1.4 reflects the situation. Let the radius *NL* of the circle with center *N* be *a*/2, the length of the tangent *LM* be *b*, then we directly realize that

$$MO = \frac{a}{2} + \sqrt{\frac{a^2}{4} + b^2},$$

from which Descartes concludes right away that this is the line segment *z* in question. From the same figure he deduces that *MP* is the solution of the equation $y^2 = -ay + b^2$. For the solution of $z^2 = az - b^2$, he uses an analogous method in another figure. The equation $z^2 = -az - b^2$ without positive root is, of course, not discussed.

After these algebraic preparations, Descartes proceeds with his basic method for analytic geometry, which he demonstrates by means of Pappus's problem mentioned above.

This problem was presented to him in 1631 by Jacobus Golius (1596–1667), professor of mathematics and Arabic in Leiden. Within several weeks Descartes sent a letter to Golius with his solution, which he also incorporated into the first book of the *Géométrie* (pp. 309–314 and 324–335). His treatment of this problem was the first application of the method that forms the basis of analytic geometry. Moreover, the result inspired him, among other things, in his views on what he called "admissible curves."

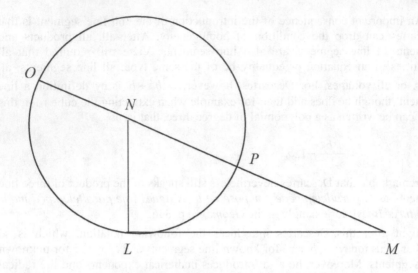

FIGURE 1.4

Prior to his solution Descartes discussed the place of this problem in antiquity, including a number of pointed jabs at the Greek mathematicians in general, and Pappus in particular.

He then announces his solution without much enthusiasm (*Géométrie*, p. 309):

> *En sorte que ie pense auoir entierement satisfait a ceque Pappus nous dit auoir esté cherché en cecy par les anciens & ie tascheray d'en mettre la demonstration en peu de mots, car il m'ennuie desia d'en tant escrir.* [So that I think that I have fully accomplished that which, according to Pappus, the ancients wanted and I will try to give the proof in few words, as it already annoys me to write so much about it.]

To clarify we first give Descartes's solution of Pappus's problem for four lines. We stay close to the original proof, including the notations. The associated Figure 1.5 is also borrowed from the *Géométrie*.

Let AB, AD, EF, and GH be four given lines in the plane. To each line corresponds a fixed direction. These directions are represented in the figure by a dotted line from a point C: CB, CD, CF, and CH. We want to find the position of the point C for which the proportion $CB.CD : CF.CH$ has a given value, which Descartes chooses to be 1. We are thus looking for the equation of a plane curve that is defined geometrically. Both Fermat and Jan de Witt proceed in a different manner: they begin with a given equation and examine which curve it describes. The method of Descartes consists of choosing the line through A and B as the abscissa-axis, with A as origin, and the direction conjugate to AB as direction of the ordinate-axis, in this case from B to C. Thus the point C has abscissa AB and ordinate BC. The first distance we want, CB, is therefore equal to y.

The angles at A are fixed, as are the angles at B, so that the angles of triangle ABR are known, and therefore also the ratios of the sides. Descartes sets

$AB : BR = z : b$, which gives $BR=bx/z$, with known b and z. In our figure, therefore, we have

$$CR = y + \frac{bx}{z}.$$

The angles of triangle DCR are also known, hence again also the ratios of the sides. Descartes sets $CR : CD = z : c$ with the same z as before. With CR as above, this give the second distance CD:

$$CD = \frac{c \cdot CR}{z} = \frac{cy}{z} + \frac{bcx}{z^2}.$$

Descartes proceeds in an analogous manner. First he sets the known distance AE in triangle SBE equal to k, so that $EB = k + x$. For the known ratio of the sides BE and BS he writes $BE : BS = z : d$, again with the same z, whence

$$BS = \frac{d \cdot BE}{z} = \frac{d(k+x)}{z}.$$

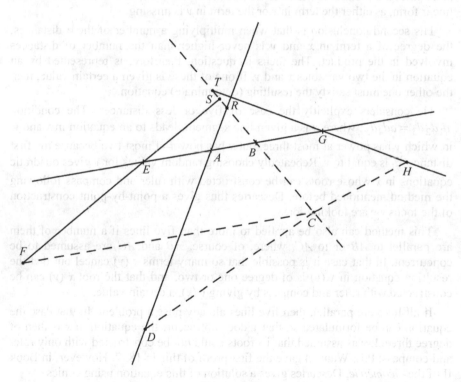

FIGURE 1.5

so that

$$CS = BC + BS = y + BS = \frac{zy + dk + dx}{z}.$$

Next Descartes sets the known ratio $CS : CF$ equal to $z : e$, which gives the following result for CF, the third distance we were looking for:

$$CF = \frac{ezy + dek + dex}{z^2}.$$

He continues in the same manner: he sets $AG = l$ (hence $BG = l - x$), $BG : BT = z : f$, $TC : CH = z : g$, and finally finds the fourth distance, CH:

$$CH = \frac{gzy + fgl - fgx}{z^2}.$$

The first conclusion Descartes draws is that, regardless of the number of given lines, the distances sought are linear forms in x and y. Of course he does not use this expression, but describes the look of such a form. For the cases where one or more lines are parallel to AB or to BC, he makes an exception to his definition of linear form, as either the term in x or the term in y is missing.

His second conclusion is that when multiplying a number of these distances, the degree of a term in x and y is never higher than the number of distances involved in the product. The locus in question, therefore, is represented by an equation in the two variables x and y. If one of these is given a certain value, then the other one must satisfy the resulting (determinate) equation.

He considers explicitly the case of five or less distances. The condition $d_1 d_2 d_3 = a d_4 d_5$, where a is a given line segment, leads to an equation in x and y, in which y has power at most three, but x has power at most two because the first distance d_1 is equal to y. Repeatedly choosing random values for y gives quadratic equations in x whose roots can be constructed with ruler and compass following the method mentioned before. Descartes thus gives a point-by-point construction of the locus we are looking for.

This method can also be applied to more than five lines if a number of them are parallel to AB or to AC, where, of course, AB and AC are assumed to be concurrent. In that case it is possible that so many terms x (y) cancel out that the resulting equation in x (y) is of degree one or two, and that the root x (y) can be constructed with ruler and compass by giving y (x), a certain value.

If all lines are parallel, then five lines already pose a problem. In that case the equation can be formulated so that y does not occur; the equation in x is then of degree three. It was assumed that its roots could not be constructed with only ruler and compass; P.L. Wantzel gave the first proof of this in 1837. However, in Book II of the *Géométrie*, Descartes gives a solution of this equation using conics.

For at most nine lines that are not all parallel, the method of Descartes gives an equation in which x has degree at most four because the first distance is equal to y.

Further on, in Book II, Descartes shows that in this case a solution can be given by intersecting conics. Analogously, for at most thirteen lines we obtain an equation of degree at most six in x. Descartes later solves this using a "higher" curve, namely the "trident" or "parabola of Descartes." We will come back to this curve in Section 5.3 of the appendix.

At this point Descartes interrupts his treatment of Pappus's problem to insert an overview of his classification of plane curves. He postpones his closer elaboration of Pappus's problem, which culminates in his conclusion that in the case of three or four lines the solution is a conic, to pp. 324–335 of the *Géométrie*, after which he treats the problem for five lines on pp. 335–341. In Section 5.2 of the appendix we will discuss this elaboration on plane curves more closely, and in Section 5.3 we will give Descartes' solution of Pappus's problem for five lines. We will now first continue Descartes' treatment of Pappus's problem for four lines.

Above, we have already deduced values for CB, CD, CF, and CH (see also Figure 1.5). The condition $CB \cdot CF = CD \cdot CH$ then gives

$$y^2(ez^3 - cgz^2) = y(cfglz - dekz^2 - dez^2x - cfgzx + bcgzx) + bcgflx - bcfgx^2.$$

If $ez^3 - cgz^2$ is negative, then we multiply both sides by -1. Descartes only considers the values of y for which C lies inside the angle DAG. After introducing suitable new coefficients m and n, which depend on b, c, d, e, f, and l, Descartes reduces the equation to the form

$$y^2 = 2my - \frac{2nxy}{z} + \frac{bcfglx - bcfglx^2}{ez^3 - cgz^2}.$$

The roots of this equation are

$$y = m - \frac{nx}{z} \pm \sqrt{\left(m^2 - \frac{2mnx}{z} + \frac{n^2x^2}{z^2} + \frac{bcfglx - bcfglx^2}{ez^3 - cgz^2} \right)}$$

of which Descartes only considers the one with the plus sign. For simplification, Descartes also introduces the quantities o and p, which are again dependent on the aforementioned coefficients, giving

$$y = m - \frac{nx}{z} + \sqrt{m^2 + ox - \frac{px^2}{m}}$$

for the root. Here, m, n, z, o, and p are known quantities.

Descartes first remarks that the locus is a line if the expression under the radical sign is zero or a perfect square. Next, he remarks straight off that in the other cases the locus is one of the three conics or a circle. For the construction of the parabola as a solution Descartes refers to the corresponding problem in the first book of the *Conica* of Apollonius. In the remaining cases he gives, without further explanation, the characteristic quantities for the curve (center, latus rectum, symmetry axes, vertices, ...) in terms of the coefficients of the equation above. For the actual construction of the curves from this he also refers to Apollonius, after which he shows that the curves constructed this way coincide with the curves that appear as solutions for Pappus's problem.

Of course, many cases can be distinguished with respect to the mutual position of the points that arise in the course of the construction, depending on the parameters of the equation. Descartes treats these in detail.

Finally, he gives a numerical example (see once more Figure 1.5). In this example the following values hold:

$$EA = 3;\ AG = 5;\ AB = BR;\ BS = BE\,/\,2;\ GB = BT;\ CD = 2CR\,/\,2;$$
$$CF = 2CS;$$
$$CH = 2CT\,/\,3;\ \angle ABR = 60°.$$

The condition $CB.CF = CD.CH$ then leads to the equation

$$y^2 = 2y - xy + 5x - x^2.$$

The locus turns out to be a circle, which Descartes shows meticulously by means of his previous considerations.

With this Descartes concludes his treatment of Pappus's problem for four lines. In Section 5.3 of the appendix we will discuss the solution of the problem for five lines.

1.6 As noted before, Fermat was inspired by the work of Apollonius. Around the end of 1635 he completed a reconstruction of two lost works of Apollonius, the *Loci Plani*, which, as the title says, deal with plane loci: that is, lines and circles. This led him to write his first and fundamental contribution to analytic geometry, entitled *Ad Locos Planos et Solidos Isagoge* [Introduction to Plane and Solid Loci], a study consisting of only eight pages, followed by a three-page appendix.

Right from the beginning his presentation is clearly different from that of Descartes. A first example of this is his notation, which he took from Viète, even though the printed version of 1679 includes numerical exponents as used by Descartes. His setting up is also different. We have already seen that Descartes took Pappus's problem for four lines as a starting point and established an equation for the sought locus, which he submitted to a precise examination with as final aim the construction (that is, the constructive solution) of equations.

Fermat went to work in the opposite manner; roughly speaking, he started with an equation in two variables and, using the properties of the conics from antiquity, he examined which curve it represented. This way he approached the problem of the loci in its most general form. Indeed, his reproach to the ancient mathematicians was that they had not tackled this problem in a general enough setting.

Fermat was the first to realize that an equation in two variables represents a curve. He states this insight in the following historical sentence:

Quoties in ultima aequalitate duae quantitates ignotae reperiuntur, fit locus loco, & terminus alterius ex illis describit lineam rectam, aut curvam, linea recta unica & simplex est, curva infinita, circulus, parabole, hyperbole, ellipsis, &c. [Whenever the final equation has two unknown quantities, the locus has a fixed position and the extremity of one of the two unknowns describes a straight or curved line; the straight line corresponds to only one type and is simple; there are infinitely many types of curved lines: a circle, a parabola, an ellipse, etc.]

The most important novelty is that in contrast to his predecessors, Fermat does not limit himself to determinate equations – that is, one equation in one variable that must be resolved, or a system with as many equations as variables, but considers indeterminate equations in two variables.

He represents the two unknown quantities in the equations as line segments in a plane. He measures off the first of the two along a fixed half-line, starting at its origin. He then sets the second one in an upward direction from the origin of the first variable under a fixed, often right, angle with the first half-line (see Figure 1.6).

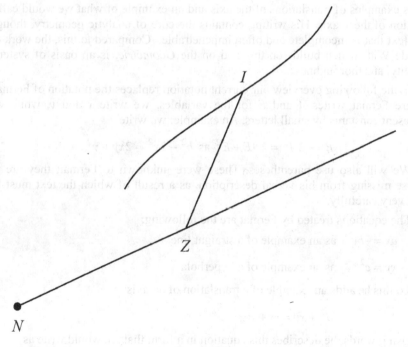

FIGURE 1.6

Thus Fermat uses a coordinate system with one axis. The *y*-axis does not occur; it comes up for the first time in a posthumous publication from 1730 by C.

Rabuel (1669–1728). From now on we will speak of the abscissa and ordinate instead of the first and second variable.

The curve in question is then generated by the extremity of the ordinate for varying abscissa. Fermat calls this extremity the *terminus localis*. Generally he only considers that part of the curve that lies in what we would call the first quadrant. We will see that Jan de Witt follows this method.

The *Isagoge* of Fermat has one central theorem:

> ...*modo neutra quantitatum ignotarum quadratum praetergrediatur, locus erit planus aut solidus, ut ex dicendis clarum fiet.* [... on condition that none of the two variables occurs in a higher power than two, the locus will be plane or solid, as will become clear from what follows.]

Here again solid locus stands for parabola, hyperbola or ellipse.

This theorem is proved by means of seven typical examples, where Fermat initially assumes the equations to be reduced (*ultima aequatio*); he had learned this reduction from the work of Viète.

In fact, with his examples Fermat studies all standard linear and quadratic equations. In this he is much more systematic than Descartes. He moreover also gives examples of translations of the axis and an example of what we would call a rotation of the *x*-axis. His writing contains the core of analytic geometry, though in a text that is incomplete and often impenetrable. Compared to this, the work of Jan de Witt, which builds on this and on the *Géométrie*, is an oasis of system, lucidity, and thoroughness.

In the following overview our current notation replaces the notation of Fermat. Where Fermat writes A and E for the variables, we write x and y, while we represent constants by small letters. An example: we write

$$B^2 - 2A^2 = 2AE + E^2 \text{ as } b^2 - 2x^2 = 2xy + y^2 .$$

We will also use parentheses. These were unknown to Fermat; they are of course missing from his verbal description, as a result of which the text must be read very carefully.

The equations treated by Fermat are the following:

i. $dx = -by$ as an example of a straight line

ii. $xy = c^2$ as an example of a hyperbola

To this he adds an example of a translation of an axis:

$$d^2 + xy = rx + sy.$$

Using words, he describes this equation in a form that we would write as

$$(x - s)(r - y) = d^2 - rs,$$

where he remarks that we now have the same form as above if we view $x–s$ and $r–y$ as "successors" of x and y. He does not speak explicitly of a new abscissa-axis.

iii. $x^2 = y^2$; $x^2 : y^2$ constant; and $(x^2 + y^2) : y^2 = $ constant as examples of pairs of lines

iv. $x^2 = dy$ and $y^2 = dx$ as examples of a parabola.

Next he reduces $b^2 - x^2 = dy$ to the form $x^2 = d(r - y)$, where $dr = b^2$, and remarks that this is the previous case if we view $r–y$ as the successor of y.

v. $b^2 - x^2 = y^2$ as an example of a circle, at least if "the angle" is a right angle.

He then reduces $b^2 - 2dx - x^2 = y^2 + 2ry$ to $p^2 - x^2 = y^2$, where x and y have replaced $x + d$ and $y + r$ and where $p^2 = b^2 + d^2 + r^2$.

vi. $(b^2 - x^2) : y^2 = $ constant as an example of an ellipse.

He adds to this that if the constant has value 1 and "the angle" is a right angle, then this is a circle, but if the angle is not right, it is indeed a true ellipse.

vii. $(x^2 - y^2) : y^2 = $ constant as an example of a hyperbola

The method Fermat uses for his proofs is essentially the same as that of Jan de Witt. Using the coefficients that occur in the equation, he describes, with words, the line or curve in question and shows that the abscissa and ordinate of an arbitrary point on it satisfy the given equation. The converse, the compositio or synthesis, is often missing or is disposed of as obvious: *est facilis compositio*. This corresponds to the proof that any point whose abscissa and ordinate satisfy the equation lies on the curve.

In his proof Fermat of course needs a characteristic property of the curve that is being discussed. For this he calls on one of the properties that Apollonius gave for the different conics, which Fermat assumes known by his readers. Thus for the ellipse he uses a property that we will need later on, which the reader can find in Note [3.5] of the translation. The corresponding figure is included here as Figure 1.7. In this CAG is the major axis of an ellipse with center A, and ED lies in the conjugate direction. The characteristic property states that the point D lies on the ellipse if and only if the ratio $DE^2 : CE \cdot EG$ is constant.

As a simple example of the method used by Fermat we choose the equation $x^2 = dy$. In this case Fermat chooses an abscissa-axis with origin N and an ordinate-axis in the conjugate direction, here ZI (see Figure 1.8). Next he describes, with words, a parabola with vertex N whose symmetry axis lies in the direction of the ordinate-axis, with conjugate axis in the direction of the abscissa-axis and latus rectum d. If PI is drawn parallel to NZ, then if the parabola goes through I, based on the characteristic for a parabola given by Apollonius, we have $PI^2 = d \cdot PN$.

Here $PI = NZ = x$ and $NP = ZI = y$, so that a point on the curve satisfies $x^2 = dy$, which is precisely the equation of the curve that we started out with.

Right away Fermat remarks that, conversely, a point I satisfying $x^2 = dy$ lies on the parabola. Once more the *compositio* is missing. In the example of the circle he is clearer on the nature of the point I; there he states explicitly that I lies on the circle in question and shows that the coordinates x and y satisfy $b^2 - x^2 = y^2$.

In addition to the examples of translations of axes mentioned above, Fermat also gives an example of a rotation of an axis, though without using this term. He announces this problem with the words:

> *Difficillima omnium aequalitatum est quando ita miscentur A^2 & E^2 ut nihilominus homogenea ab A in E afficiantur una cum datis &.* [The most difficult of all equations is that where x^2 and y^2 occur in such a way that there are also terms with xy and constants...]

His example is the equation

$$B^2 - 2A^2 = 2AE + E^2$$

for us $\qquad b^2 - 2x^2 = 2xy + y^2$, that is, $b^2 - x^2 = (x+y)^2$.

We will follow Fermat's solution closely, though in general using our own notation.

Fermat first chooses an abscissa-axis with origin N and with ordinate direction perpendicular to it (see Figure 1.9). V is an arbitrary point whose abscissa NZ (= x) and ordinate ZV (= y) satisfy the equation mentioned above. We are interested in the locus of V for variable Z.

To find this Fermat describes a circle with center N and radius b that meets the perpendicular through Z at I and the abscissa-axis at M.

From Figure 1.9 it becomes clear that

$$NM^2 - NZ^2 = ZI^2 = (ZV + VI)^2 \qquad \text{(i)}$$

and from the given equation follows

$$NM^2 - NZ^2 = (ZV + NZ)^2,$$

so that $\qquad VI = NZ (= x)$.

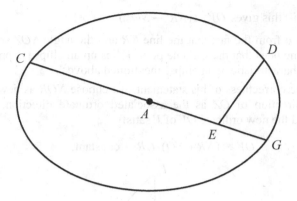

FIGURE 1.7

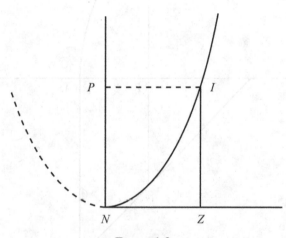

FIGURE 1.8

Then Fermat draws the line segment MR parallel to IZ and equal to NM. Let O be the intersection point of NR and the extension of IZ. Clearly $NZ=ZO$; as we have already seen that $NZ = VI$, this gives $OV = ZI$. From (i) then follows

$$NM^2 - NZ^2 = OV^2. \qquad\qquad \text{(ii)}$$

Fermat now remarks that the ratios $NM^2 : NR^2$ and $NZ^2 : NO^2$ are "given," without mentioning that they are equal, which is essential here. He does not use that their common value is equal to 1:2.

From $NM^2 : NR^2 = NZ^2 : NO^2 = \text{constant},$

follows $(NM^2 - NZ^2) : (NR^2 - NO^2) = \text{constant}.$

Together with (ii) this gives $OV^2 : (NR^2 - NO^2)$ = constant.   (iii)

From this and from the fact that the line NR and the angle NOZ are fixed he immediately concludes that the variable point V lies on an ellipse. Apparently he calls upon the characteristic of an ellipse mentioned above.

To check the correctness of his statement, we choose NOR as new abscissa-axis and the direction of OZ as the associated ordinate direction. The new abscissa NO and the new ordinate OV of V satisfy

$$OV^2 : (NR + NO) \cdot OR = \text{constant.}$$

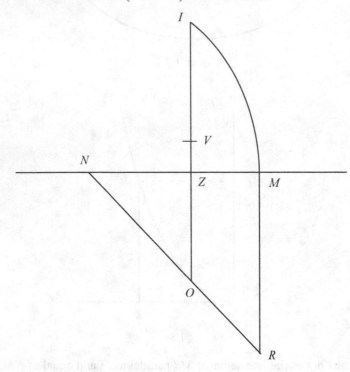

FIGURE 1.9

Based on the characteristic property mentioned above, V lies on an ellipse with center N, symmetry axis NR, and associated conjugate axis parallel to OV, and hence to RM.

All this can be verified through a simple calculation. If we set $NO = u$ and $OV = v$, then $v^2 : (2b^2 - u^2) = 1:2$ follows from (iii) if we consider that $NM^2 : NR^2 = 1:2$.

This means that

$$\frac{u^2}{2b^2} + \frac{v^2}{b^2} = 1,$$

which is exactly the equation of an ellipse with center N and axes of length $2b\sqrt{2}$ and $2b$.

As $u = x\sqrt{2}$ and $v = x + y$, we have, as should be, $b^2 - x^2 = (x+y)^2$.

Fermat notes that all cases with a mixed term xy can be treated with similar methods.

As the crowning glory (*coronidis loco*) of this work Fermat adds the following *propositio*. Consider arbitrarily many given lines in the plane and a point with line segments drawn towards these lines under given angles, then the points for which the sum of the squares of these line segments equals a given area lie on a conic.

One can see this problem as a variant of Pappus's problem, with sums instead of products. Fermat does not solve this problem but a simplified version of it: given two points M and N, determine the locus of the points I such that $IM^2 + IN^2$ has a given ratio to the area of triangle IMN. This locus proves to be a circle, of which he gives an equation, followed by a construction independent of this.

Not without conceit Fermat concludes his *Isagoge* with the following remark:

> If this discovery had preceded the two books on plane loci that I recently reconstructed, then the proofs of the theorems concerning loci would certainly have been more elegant.

1.7 Fermat added a three-page appendix to his *Isagoge*, entitled *APPENDIX AD ISAGOGEM TOPICAM continens solutionem problematum solidorum per locos.* [Appendix to the introduction to the loci, containing the solution of spatial problems by means of loci.]

In this appendix Fermat applies his theory to solving algebraic equations in one variable, the so-called determinate equations. His technique consists of introducing a cleverly chosen second variable in order to transform the problem into determining the intersection point of two plane curves. He limits himself to equations of degree three or four and shows that conics suffice for solving these.

His first example is the equation

$$x^3 + bx^2 = c^2 b.$$

He introduces a new variable y by setting

$$x^3 + bx^2 = c^2 b = bxy,$$

which leads to the system

$$x^2 + bx = by \qquad\qquad\qquad \text{(i)}$$

$$c^2 = xy. \tag{ii}$$

Using this, the variable x can be found as the abscissa of an intersection point of the parabola with equation (i) and the hyperbola with equation (ii). As Fermat only considers one "quadrant," he finds only one intersection point. He is not interested in more anyway. Any further examination of this root, such as checking whether this root indeed satisfies the equation and whether there are more roots, such as $x=0$, is missing. He again dismissed these matters with the words *est facilis ab analysi ad synthesim regressus*.

Fermat notes that all cubic equations can be solved using a similar method.

The second example is of degree four:

$$x^4 + b^3 x + c^2 x^4 = d^4,$$

which he rewrites as

$$x^4 = d^4 - b^3 x - c^2 x^2.$$

By setting both sides equal to $c^2 y^2$ one easily sees that the solution is found by intersecting the parabola with equation $x^2 = cy$ and the circle with equation $c^2 y^2 = d^4 - b^3 x - c^2 x^2$. Both here and elsewhere in the appendix Fermat mentions explicitly that one already knew how to eliminate the term x^3 in a degree-four equation from the work of Viète, so that his example has a general validity.

After this Fermat brings up an old problem: determining x, the greatest of the two geometric means of b and a, where $b > a$. He of course assumes known that this satisfies the equation $x^3 = b^2 d$. To solve this he sets $x^3 = bxy$ and $b^2 d = bxy$. The x in question is therefore determined by an intersection point of the parabola with equation $x^2 = by$ and the hyperbola with equation $xy = bd$. Fermat gives a geometric explanation for this. He also gives a solution with two parabolas. For this he replaces $x^3 = b^2 d$ by $x^4 = b^2 dx$ after which he sets $x^4 = b^2 y^2$ and $b^2 dx = b^2 y^2$, so that the solution follows from "the" intersection point of the parabolas with equations $x^2 = by$ and $y^2 = dx$. He mentions that this method can also be found in the commentary of Eutocius (ca 480 AD) to Archimedes.

With a reproach to Viète, Fermat gives an application of his "elegant method" to solving the general degree-four equation, namely by using the intersection of a parabola and a circle. He objects to Viète's use of a cubic equation, the resultant, when solving a degree-four equation. Fermat gives two characteristic examples:

$$x^4 = c^3 x + d^4 \text{ and } x^4 = c^2 x^2 - c^3 d.$$

Here too he notes that since Viète one knows how to get rid of a possible cubic term, so that these two types of equation suffice.

In the first case he completes the square on the left-hand side to the form $(x^2 - b^2)^2$, where b must still be determined, so that the equation becomes

$$(x^2 - b^2)^2 = c^3 x + d^4 + b^4 - 2b^2 x^2.$$

He then sets both sides equal to $n^2 y^2$ with $n^2 = 2b^2$ and chooses

$$x^2 - b^2 = ny. \tag{i}$$

We moreover have

$$c^3 x + d^4 + b^4 - 2b^2 x^2 = n^2 y^2, \tag{ii}$$

so that the variable x is determined by an intersection point of a parabola and a circle with equations (i) and (ii) respectively.

The second case is more complicated. As before he completes both members to

$$(x^2 - b^2)^2 = b^4 - 2b^2 x^2 + c^2 x^2 - c^3 d,$$

and sets both equal to $n^2 y^2$. Here b and n must still be determined. Again he chooses

$$x^2 - b^2 = ny.$$

We also have $b^4 - 2b^2 x^2 + c^2 x^2 - c^3 d = n^2 y^2,$

that is: $(c^2 - 2b^2)x^2 + b^4 - c^3 d = n^2 y^2.$

It is now clear that in order to obtain a circle, we must choose b in such a way that $2b^2 > c^2$ and n in such a way that $n^2 = 2b^2 - c^2$.

The choice $2b^2 < c^2$ would lead to a hyperbola instead of the desired circle, which is simpler for a constructive solution. Fermat disregards the choice $2b^2 = c^2$.

Finally, Fermat shows that the problem of the two geometric means can also be solved in this manner. Here is a sketch of his method. First he again replaces $x^3 = b^2 d$ by $x^4 = b^2 dx$ and, as above, switches over to

$$(x^2 - b^2)^2 = b^4 + b^2 dx - 2b^2 x^2 = n^2 y^2,$$

where $n^2 = 2b^2$, so that x is the abscissa of the intersection point of the parabola with equation

$$x^2 - b^2 = ny$$

and the circle with equation

$$b^4 + b^2 dx - n^2 x^2 = n^2 y^2 ; \text{ that is: } b^2 + dx - 2x^2 = 2y^2.$$

His conclusion is: whoever sees this will try in vain to solve the problem of the mesolabium, the trisection of the angle and similar problems with the help of plane curves, that is, using lines and circles. A mesolabium is an instrument for constructing the geometric means of two line segments, invented by Eratosthenes (ca 276 – ca 195 BC). See [65].

1.8 The destiny of Fermat's *Isagoge* differed in many respects from that of the *Géométrie* of Descartes.

Fermat lived isolated as Royal Counselor and Commissioner of appeals of the parliament of Toulouse and later rose through the ranks, but never ventured far from the city. He never even visited Paris. In addition to his administrative functions he practiced mathematics intensively, though he continued to consider himself an outsider. His contacts with the mathematical world were mainly through letters, among which a prominent place was taken by the correspondence with members of a Parisian group of mathematicians led by Etienne Pascal, the father of Blaise. Père Marin Mersenne managed the extensive correspondence of this group and determined who the right person for the various subjects was.

Through this group, Fermat made his *Isagoge* (with Appendix) known to the world, and it is also thus that Descartes and Frans van Schooten Jr. obtained copies of it. Fermat, however, did not authorize the publication of his *Isagoge*, something which applied to virtually all of his work. Let us note that already prior to 1650, the brothers Elsevier had plans to publish the work of Fermat, but these were never realized. Only in 1679, fourteen years after his death, did his son Clément-Samuel provide a first printed version of his mathematical works under the title *Varia Opera Mathematica D. Petri de Fermat, Senatoris Tolosani*. However, at that point his work had already been surpassed by the *Géométrie* of Descartes and the work of his followers.

1.9 This *Géométrie*, which appeared directly in print in 1637 through Jan Maire, a bookseller in Leiden, was initially not appreciated by everyone. The reason for this was not only the inaccessible style of this work and the French language in which it was written, but also the fact that many continued to prefer the synthetic method of the Greeks. It was important, therefore, that there be mathematicians who ensured the diffusion of the ideas laid down in the *Géométrie*.

Among them, Frans van Schooten Jr. deserves a special mention. He made the *Géométrie* accessible to all scholars through his translation of it into Latin, the lingua franca of the scientific world, and through his lucid explanations and clear drawings. For the history of this work the reader is referred to the introduction to the translation of *Liber Primus*.

In France, G.P. de Roberval (1602–1675) spent much time on introducing the method of Descartes. Although an important work of his concerning the formulation of equations of loci and the resolution of equations using the intersection of plane curves only appeared after his death, he presumably introduced his students to the analytic geometry of Descartes during his lectures at the Collège de France.

In 1639, the *Notae Breves*, a commentary on the *Géométrie* written by Florimond Debeaune (1601–1652), came out. In this, the *Géométrie* is followed painstakingly, commented on exhaustively, and completed in a systematic manner, in particular where the classification of quadratic equations in two variables is concerned. Van Schooten Jr. included these *Notae Breves* in all editions of his *Geometria à Renato Descartes, etc.* From the second edition on, this work also included a posthumous treatise of Debeaune on algebraic equations.

Philippe de la Hire (1640–1718) was one of the French who later gave overviews and explanations of the *Géométrie*. After two dissertations in synthetic style concerning conics, in 1679 he published a work in three parts of which the first part gives a planimetric treatment of conics as plane curves defined by their focal properties. The second part, *Les Lieux Géométriques*, is written in the style of *Liber Secundus* of Jan de Witt, but it also contains an introduction to analytic geometry in three or more dimensions. De la Hire was the first to give a simple example of a surface as locus, represented by a quadratic equation in three variables. Fermat and Descartes had only alluded to this possibility. The first systematic treatment of analytic geometry in dimension three was by Antoine Parent (1666–1716), who submitted a paper on this to the Académie des Sciences in 1700. The third part of the work of de la Hire treats the constructive resolution of algebraic equations by intersecting plane curves.

It was John Wallis (1616–1703) who propagated the ideas of Descartes in Great Britain, though not through a translation of or commentary on the *Géométrie*, but through an original work, the *Tractatus de sectionibus conicis*, published in 1655.[1] In our introduction to *Liber Primus* we have already mentioned that he defined conics by their equations. Of course, these equations did not appear out of nowhere: they were inspired by the symptomata of Apollonius. In another respect Wallis also advanced the algebraicization: for him coordinates were no longer line segments, but numbers. He also introduced negative abscissa and ordinates. Furthermore, in his *Arithmetica Infinitorum*, also from 1655, he linked analytic geometry and infinitesimal methods.

Nevertheless, the ideas of Wallis did not catch on with all English mathematicians. Some felt more at home with the synthetic methods from antiquity; for example Newton's teacher, Isaac Barrow (1630–1677), was strongly opposed to this new method. This is in contrast to Newton, who valued the *Géométrie* greatly. Wallis also made enemies in continental Europe, partly due to his claiming certain priorities. According to him, the *Géométrie* was based on the *Artis analyticae praxis* from 1631 by Thomas Harriot (1560–1621) and that the

---

[1] Editors note. One of the referees has pointed out that it can be argued that Wallis's work was almost entirely independent of Descartes's work. We refer to the English translation of *Arithmetica Infinitorum, The arithmetic of infinitesimals* by Jacqueline A Stedall, Sources & Studies in the History of Mathematics & the Physical Sciences, 2004.

Lieux Géométriques of de la Hire was a plagiarism of *De sectionibus conicis* of Wallis.

Of course the work of Jan de Witt has frequently been compared to *De sectionibus conicis*; we already mentioned this in the introduction to the translation of *Liber Primus*. Based on what we have said earlier, de Witt cannot have been influenced by Wallis; after all, the latter defined conics by their equations, whereas de Witt based himself on a kinematic manner of generation and used this to show which equation corresponded to which curve. Moreover, coordinates are numbers for Wallis while for de Witt they are still line segments. One might say that de Witt is more conservative in his approach than Wallis. Finally, the first draft of de Witt's work was ready in 1649, six years before the publication of Wallis's *De sectionibus conicis*.

Let us also name Wallis's *Treatise of Algebra, both Historical and Practical*, published in 1685. In it he considered the constructive resolution of equations. It also includes the famous "Wallis conocuneus" (conical wedge, a figure with a circular base like a cone, but having a ridge or edge instead of the apex).

Finally, let us mention Jacques Ozonam (1640–1717), who in 1687 wrote a treatise, ordered following the gradually accepted layout: first the geometry of conics, then loci and finally constructions of roots of algebraic equations.

By that time, however, the interest in analytic geometry was reduced, at least temporarily, by the rise of differential and integral calculus: Leibniz with his *Nova Methodus* of 1684 [44], a journal article of only six pages, and Newton, whose fundamental contribution to this new subject had been circulating amongst colleagues in manuscript form since 1665 before coming out in print in 1704.

1.10 To conclude we will briefly present the structure of *Liber Secundus* and sketch Jan de Witt's method. The reader will find a detailed overview of the contents in the summary.

The core of the book comes down to the following. First, Jan de Witt gives equations in x and y for the straight line and the conics, in standard form with respect to a coordinate system chosen by him, which we will come back to later. These correspond to the following well-known forms:

(i) $\qquad\qquad y = ax + b, \; x = ay + b$ $\qquad\qquad\qquad\qquad$ (line);

(ii) $\qquad\qquad y^2 = ax + b, \; x^2 = ay + b$ $\qquad\qquad\qquad\qquad$ (parabola);

(iii) $\qquad\quad a^2 x^2 - b^2 y^2 = 1, \; a^2 y^2 - b^2 x^2 = 1, \; xy = c = 1$ $\qquad$ (hyperbola);

(iv) $\qquad\quad a^2 x^2 + b^2 y^2 = 1$ $\qquad\qquad\qquad\qquad\qquad\qquad$ (ellipse).

The restriction to positive coefficients and the homogeneity condition force him to distinguish more cases and to give other forms, which formally differ from these. These can be found in the summary.

He then shows that these standard forms indeed represent the curves in question. That is to say, the coordinates of every point on such a curve satisfy the corresponding equation. The converse, that every point of which the coordinates satisfy the equation indeed lies on the corresponding curve, is seldom shown.

After a short treatment of the straight line (which is hardly considered by others), all attention is on the conics. Three statements, which Jan de Witt announces as *Regula Universalis* (Universal Rule), are central.

The first concerns the parabola, and states that every quadratic equation in two variables x and y of which only one occurs as a square can be reduced to one of the forms in (ii); that is, to one of the variants he gives. He also tells how this can be done: by splitting off a perfect square. He does not prove this in its full generality, but illustrates the method by means of thirteen well-chosen examples, which all correspond to rotations and to parallel translations of the abscissa-axis.

For the second *Regula Universalis* Jan de Witt assumes that the equation that is being studied can also have terms with x^2, y^2, xy, x, and y, and that it does not represent a parabola. Such an equation, he says, can be reduced to one of the forms in (iii) and (iv). Again he shows how to do this. It is now a matter not only of splitting off a perfect square, but also of a technique that corresponds to "removing the brackets," though he does not use brackets. For example, for the expression $xy + ay$, he introduces the new variable $v = x + a$ and substitutes $x = v - a$ in the expression, which transforms it into vy. He demonstrates this method with four fairly general examples.

In the third *Regula Universalis*, Jan de Witt considers the general quadratic equation in two variables in order to show that this can always be reduced to one of the forms stated above. For this he first repeats the standard forms in x and y stated before, which represent a straight line or a conic. Then he introduces new forms for conics by successively replacing the variables y and x in the forms given in (ii) – (iv) by z and/or v, where

$$z = y + px + q \text{ and } v = x + h$$

or $z = y + h$ and $v = x + py + q$.

In an obvious way this gives rise to the new forms. For example the form

$$a^2x^2 + b^2y^2 = 1$$

gives rise to

$$a^2x^2 + b^2z^2 = 1, \ a^2v^2 + b^2y^2 = 1, \text{ and } a^2v^2 + b^2z^2 = 1,$$

where z and v are as above.

The form $xy = c$ gives rise to $xz = c, \ vy = c$ and $vz = c$, but in this case Jan de Witt restricts himself to $z = y + h$ and $v = x + k$.

The third *Regula Universalis* states that the general quadratic equation in two variables can be reduced to one of the forms in (ii)–(iv) or to one of the forms deduced from these as above. The summary contains an overview of these. Strictly speaking, Jan de Witt does not prove that this rule holds. He goes to work conversely, as follows. He considers each of the new forms in z and/or v and shows in detail that for all possibilities for z and v it represents a conic, which he describes in detail with words. He does not write down the curves themselves; rather, he writes their vertices, centers and axes. However, he does not mention how to reduce a general quadratic equation to such a form, saying only *Methodo jam explicata*, which means "in the manner that has already been explained." Apparently he means the many explicit examples, in which different manners of reduction have been shown. For example on p. [283] he clearly shows the technique that is meant here. Consequently, the third *Regula Universalis* is not followed by any examples.

Like Fermat, Jan de Witt starts out with a given linear or quadratic equation in two variables and proves that it represents one of the curves in question, as discussed above. Like Fermat, he uses a coordinate system with one axis. The method is described in Section 1.6 of this introduction (see also Figure 1.6). Jan de Witt sets out the coordinates in the same direction as Fermat, as a result of which he too restricts himself to the "first quadrant." For this de Witt uses a fixed phrase, which he repeats time after time:

> Let A be the immutable initial point and let us suppose that x extends indefinitely along the straight line AB, and let the given or chosen angle be equal to angle ABC.

The word "indefinite" requires an explanation. It is the same word as that used in the Latin text, *indefinite*. That word can also be read as "infinite" or "indeterminate." "Infinite" could be confusing. The quantity x, like y, remains finite, as both represent line segments. Its length can take on arbitrary values, without restriction, but remains finite. Sometimes in the translation, the word "indefinite" is replaced by "indeterminate" or "arbitrary."

In the proof that a given equation represents a certain curve (as mentioned above), de Witt proceeds in a standard manner. Using the coefficients that occur in the equation, he describes the curve that he has in mind using words. The quantities that determine this curve, such as latus rectum, vertex, symmetry axis and such, seem to appear out of nowhere, but were anticipated by him in a clever way. He calls this part of the proof the *determinatio* or the *descriptio*. After this he takes an arbitrary point on the curve with abscissa x and ordinate y. Using the geometric characteristics that he formulated in *Liber Primus* for the straight line and each of the conics, he then shows that the coordinates of the point chosen by him indeed satisfy the given equation. From p. [260] on, he explicitly calls this part of the proof the *demonstratio*.

In general, de Witt does not include the converse, the *compositio* or *synthesis*. This corresponds to the proof that every point whose coordinates satisfy the equation indeed lies on the curve that is being considered. As a result of this he misses the second branch of the hyperbola. Remarkably, he pays no attention to degenerations.

Like *Liber Primus*, *Liber Secundus* is a tightly ordered text. The central elements are the *Regulae Universalis* mentioned above, supported by fourteen *Theoremata*, many *Exempla* and three *Problemata*.

A *Theorema* consists of a *propositio* possibly followed by a number of *corollaria*. In the *propositio* a theorem is stated and proved; the *corollaria* give consequences of the *propositio*.

A *Problema* also consists of a propositio and possible *corollaria*. In this case the propositio states and resolves a constructive problem; the *corollaria* give further properties of the constructed figure.

The *propositiones* are numbered consecutively throughout the whole work, whether they belong to a *Theorema* or to a *Problema*.

As in *Liber Primus*, Jan de Witt has added four types of marginal notes to the text. In three of the four cases he refers to these by means of superscript numbers in the text. In the translation these marginal notes are incorporated as footnotes. This concerns:

i. References to the *Elements* of Euclid. These have the same standard form as in *Liber Primus*. For example, "per 16 secti" refers to Theorem 16 of the sixth book of the *Elements*. In the footnote in the translation this is denoted by "VI, 16."

ii. References to *Elementa Curvarum Linearum* itself. For example, "per 1 primi hujus" refers to Proposition 1 of *Liber Primus*. To make it easier to find these we have added the number of the corresponding page of the Latin text. The footnote then reads "Prop. 1, Lib. I, p. [162]." Likewise "per 3 Corol. 6 primi hujus" becomes "Corollary 3 of Prop. 6, Lib. I, p. [191]."

iii. Technical clarifications of proofs in the text. Again an example (p. [302]): as a footnote the marginal comment "quippe quadr. ex *HO* aequatur *GAF* rectang. ex hypoth." becomes "because, by hypothesis, the square on *HO* is equal to the rectangle *GAF*."

There are also marginal notes that are not referred to by a number. In general this concerns the name of a special case, such as on p. [318]: "Casus 1^{mus}, cùm Locus est Hyperbola." In the translation such a marginal note has in general been incorporated into the text as a heading, here as: "First case where the locus is a hyperbola."

In one case (p. [305]) the comment concerns the definition of a symbol used by Jan de Witt; it can be found in Note [4.2].

Summary

In this summary the theorems and their corollaries (*corollaria*) are restated in our modern notation, without proofs.

The essence of the statements, however, has been preserved. For the proofs we refer to the text, the translation, and the notes. This summary aims only at giving a global survey of the contents of the book.

For an overview of the structure of *Liber Secundus* and the method followed by Jan de Witt, we refer the reader to the Introduction, Section 1.10.

Chapter I

In this chapter Jan de Witt examines linear equations in x and y. Here x and y are clearly interpreted as line segments, as is evident, among other things, from the wording "the initial point of one of the quantities." The coefficients that occur are consequently always positive.

As in *Liber Primus*, the curves that are represented by the equations are called "loci."

The constructions in this work involve an x-axis, but no y-axis occurs. This x-axis is introduced as follows: Jan de Witt takes a fixed point A on an arbitrary line and chooses this point as the fixed and immutable initial point of the variable x, which can extend indefinitely along this fixed line through A (always to the right). A given x is then represented by a line segment on this line, for example AE. From the endpoint of this x a line segment y is drawn "upward" under a fixed given or chosen angle (see ED in Figure 2.1). We would say that he restricts himself to the first quadrant. As was noted in the Introduction, the y-axis was only introduced towards the end of the 17th century, by Claude Rabuel (1669–1728).

Chapter I consists of the following theorems:

Theorem I. *Proposition* 1.

If the equation is $y = bx / a$, then the required locus will be a straight line.

Theorem II. *Proposition* 2.

A.W. Grootendorst et al. (eds.), *Jan de Witt's Elementa Curvarum Linearum*,
Sources and Studies in the History of Mathematics and Physical Sciences,
DOI 10.1007/978-0-85729-142-4_2, © Springer-Verlag London Limited 2010

If the equation is $y = bx/a + c$, then the required locus will be a straight line.

Theorem III. *Proposition* 3.

If the equation is $y = bx/a - c$, then the required locus will be a straight line.

Theorem IV. *Proposition* 4.

If the equation is $y = c - bx/a$, then the required locus will be a straight line.

Theorem V. *Proposition* 5.

If the equation is $y = c$, then the required locus will be a straight line.

Theorem VI. *Proposition* 6.

If the equation is $x = c$, then the required locus will be a straight line.

Chapter II

In this chapter the following equations are considered:

I. $y^2 = ax$ or conversely $ay = x^2$;

II. $y^2 = ax + b^2$ or conversely $ay + b^2 = x^2$;

III. $y^2 = ax - b^2$ or conversely $ay - b^2 = x^2$;

IV. $y^2 = -ax + b^2$ or conversely $b^2 - ay = x^2$.

First the following theorems are proven:

Theorem VII. *Proposition* 7.

If the equation is $y^2 = ax$ or conversely $ay = x^2$, then the required locus will be a parabola

This is a direct consequence of Theorem I of *Liber Primus* (p. [162]), where this property is deduced as a characteristic (*symptoma*) of the parabola.

Theorem VIII. *Proposition* 8.

If the equation is $y^2 = ax + b^2$ or conversely $ay + b^2 = x^2$, then the required locus will be a parabola.

Theorem IX. *Proposition* 9.

If the equation is $y^2 = ax - b^2$ or conversely $ay - b^2 = x^2$, then the required locus will be a parabola.

Theorem X. *Proposition* 10.

If the equation is $y^2 = -ax + b^2$ or conversely $b^2 - ay = x^2$, then the required locus will be a parabola.

The "converse" forms (*conversim*) in Theorems VII to X are deduced by interchanging the roles of x and y in the figures and argumentations. Here, too, only the parts of the curves that lie to the right of A and above the x-axis are considered.

General Rule and method of reducing all equations that result from a suitable operation (when the required locus is a parabola) to one of the four cases that have been explained in the four preceding theorems

This general Rule gives a method of reducing the equation in question, when it represents a parabola, to one of the forms of Theorem VII, VIII, IX, or X. Jan de Witt does not mention how one first determines that the equation represents a parabola.

This method amounts to splitting off a square: if in addition to the term with x^2, the terms $\pm 2ax$, $\pm 2xy$, $\pm 2axy$ also occur, then we introduce a new variable z with respectively

$$z = x \pm a, \quad z = x \pm y \text{ or } z = x \pm ay.$$

Of course the sign used for z is precisely the sign of the corresponding term if it occurs on the same side of the equal sign as x^2. The same holds, mutatis mutandis, if y^2 is concerned.

The examples with which Jan de Witt illustrates his method are obviously chosen so that they represent parabolas.

Examples of the reduction of equations to the form of Theorem VII

1. Reduction of the equation

$$y^2 + 2ay = bx - a^2$$

to the form $z^2 = bx$,

where $z = y + a$.

This reduction is followed by the construction of the corresponding parabola (*determinatio*, also *descriptio*) and the proof that the coordinates of the points on it indeed satisfy the equation (*demonstratio*). Henceforth, in the interest of conciseness, we will refer to this procedure as *determinatio* and *demonstratio*.

2. Reduction of the equation

$$y^2 - 2ay = bx - a^2$$

to the form $z^2 = bx$,

where $z = y - a$.

For the rest of the proof we refer to the previous example.

3. Reduction of the equation

$$by - a^2 = x^2 + 2ax$$

to the form $by = v^2,$

where $v = x + a.$

Description of the construction of this parabola in the plane, followed by the *demonstratio*.

4. Statement that the equation

$$by - a^2 = x^2 - 2ax$$

can be treated analogously.

5. Reduction of the equation

$$y^2 + \frac{2bxy}{a} + 2cy = bx - \frac{b^2 x^2}{a^2} - c^2$$

to the form $z^2 = dx,$

where $z = y + \dfrac{bx}{a} + c$ and $d = \dfrac{2bc}{a} + b.$

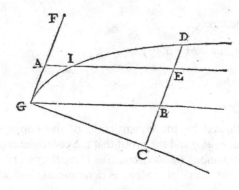

FIGURE 2.1

For the construction of this parabola (Figure 2.1), the point $G(0, -c)$ is chosen as vertex. The axis GC goes through G and lies so that $GB : BC = a : b$, where BC lies in the direction of the ordinate-axis.

Here too a *demonstratio* follows, in which Jan de Witt remarks that if the curve did not meet the x-axis, the solution would also be a parabola, but would not be constructible in a "satisfying manner" (*quod nulla tamen quaestioni satisfaciens describi possit*). By this he means that it would not lie above the x-axis. See the corresponding notes to the translation.

Note that this example was chosen very carefully.

6. The position of the curve defined by the equation

$$y^2 - \frac{2bxy}{a} - 2cy = bx - \frac{b^2x^2}{a^2} - c^2$$

is compared with that of the parabola considered in (5).

7. Reduction of the equation

$$by - \frac{b^2y^2}{a^2} - c^2 = x^2 + \frac{2bxy}{a} + 2cx$$

to the form

$$dy = v^2,$$

where

$$v = x + \frac{by}{a} + c \text{ and } d = \frac{2bc}{a} + b.$$

Note that this situation is the "converse" of that in (5). Again a complete *determinatio* and *demonstratio* follow.

Examples of the reduction of equations to the form of Theorem VIII

1. Reduction of the equation

$$y^2 - \frac{bxy}{a} = -\frac{b^2x^2}{4a^2} + bx + d^2$$

to the form

$$z^2 = bx + d^2,$$

where

$$z = y - \frac{bx}{2a}.$$

Description (*descriptio*) of the construction of this parabola in the plane, including the proof (*demonstratio*) that the curve described this way is the parabola determined by the equation. Attention is also drawn to the "converse" parabola, whose equation is obtained by interchanging the variables x and y.

2. Reduction of the equation

$$\frac{bcy}{a} + by - \frac{b^2y^2}{a^2} + \frac{c^2}{4} = x^2 + \frac{2byx}{a} - cx$$

to the form

$$by + \frac{c^2}{2} = v^2,$$

where

$$v = x + \frac{by}{u} - \frac{c}{2}.$$

Description of the construction of this parabola in the plane, including the proof that the curve described this way is the parabola determined by the equation.

Example of the reduction of equations to the form of Theorem IX

Reduction of the equation

$$y^2 + \frac{bxy}{a} - cy = ax - \frac{b^2 x^2}{4a^2} - c^2$$

to the form

$$z^2 = dx - \frac{3}{4}c^2,$$

where

$$z = y + \frac{bx}{2a} - \frac{c}{2} \text{ and } d = a - \frac{bc}{2a}.$$

Description of the construction of this parabola in the plane, including the proof that the curve described this way is the parabola determined by the equation.

Examples of the reduction of equations to the form of Theorem X

1. Reduction of the equation

$$ay - y^2 = bx$$

to the form

$$z^2 = \frac{a^2}{4} - bx,$$

where

$$z = y - \frac{a}{2}.$$

Description of the construction of this parabola in the plane, including the proof that the curve described this way is the parabola determined by the equation.

2. Reduction of the equation

$$\frac{b^2 y^2}{a^2} + dy - c^2 = \frac{2byx}{a} - x^2$$

to the form

$$c^2 - dy = v^2,$$

where

$$v = x - \frac{by}{a}.$$

Description of the construction of this parabola in the plane, including the proof that the curve described this way is the parabola determined by the equation. Another determination of the corresponding diameter and latus rectum is also given.

Problem I. *Proposition* 11.

Given a point and a line, determine the locus of all points in the plane passing through both that are equidistant from this point and line. Construct this locus.

The locus turns out to be a parabola. The term *focus* or umbilical point (*umbilicus*) is introduced here. The term *directrix* is not used yet.

Corollary 1. The line segment from a point D on a parabola to the focus is equal to the line segment from the projection of D on the symmetry axis to the vertex plus one fourth of the latus rectum. See Figure 2.2.

Corollary 2. The angle between the line segment from a point on the parabola to the focus and the tangent to the parabola at this point is equal to the angle between this tangent and the symmetry axis. This tangent bisects the angle between the first line segment and the line parallel to the axis and through the point on the parabola.

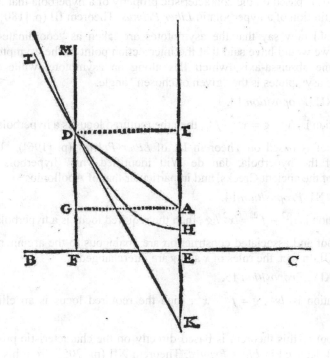

FIGURE 2.2

Chapter III

In this chapter the following equations are considered:

I. $$yx = f^2;$$

II. $$\frac{ly^2}{g} = x^2 - f^2;$$

III. $$y^2 - f^2 = \frac{lx^2}{g};$$

IV. $$\frac{ly^2}{g} = f^2 - x^2.$$

Theorem XI. *Proposition* 12.

If the equation is $yx = f^2$, then the required locus is a hyperbola.

The proof is based on the characteristic property of a hyperbola that is deduced from the definition of a hyperbola in *Liber Primus*, Theorem III (p. [180]).

We would now say that the asymptotes are taken as "coordinate-axes." In those days we would have said that the intersection point of the asymptotes is the origin of the abscissa-axis, which lies along an asymptote, while the angle between the asymptotes is the "given or chosen" angle.

Theorem XII. *Proposition* 13.

If the equation is $ly^2 / g = x^2 - f^2$, then the required locus is a hyperbola.

The proof is based on Theorem IX of *Liber Primus* (p. [196]). Using this property of the hyperbola, Jan de Witt identified "his" hyperbola with the hyperbola of the ancient Greeks, and in particular that of Apollonius.

Theorem XIII. *Proposition* 14.

If the equation is $y^2 - f^2 = lx^2 / g$, then the required locus is a hyperbola.

The proof and associated construction are analogous to the argumentation for Theorem XII. In fact, the roles of x and y are interchanged.

Theorem XIV. *Proposition* 15.

If the equation is $ly^2 / g = f^2 - x^2$, then the required locus is an ellipse (or a circle).

The proof of this theorem is based directly on the characteristic property that Jan de Witt deduced in *Liber Primus*, Theorem XII (p. [205]) from his definition of the ellipse. Of course the curve is a circle if $l = g$ and the "given or chosen" angle is a right angle.

General Rule and method of reducing all equations that result from a suitable operation (when the locus is a hyperbola or an ellipse or a circle) to one of the four cases that have been explained in the four preceding theorems

This general Rule states, without proof, that an equation of degree 2 that contains one or more of the terms xy, x^2, y^2, ax, and by can be reduced to one of the forms of Theorems XI to XIV, at least if it represents a hyperbola, an ellipse, or a circle.

The method amounts to replacing the combinations

$$xy \pm ay \text{ and } xy \pm ax$$

by vy and vx respectively, where $v = x \pm a$ and $v = y \pm a$, and splitting off a square from the combinations

$$x^2 \pm 2\,ax \text{ and } y^2 \pm 2\,ay$$

by introducing the new variables

$$v = x \pm a \text{ and } v = y \pm a.$$

As Jan de Witt did not have any parentheses at his disposal, from our point of view these computations are rather long-winded. For more details see the notes to the translation of the passage in question.

Example of the reduction of equations to the form of Theorem XI

Reduction of the equation

$$yx - cx + hy = e^2$$

to the form $lzv = f^2$,

where $z = y - c$, $v = x + h$, and $f^2 = e^2 - ch$.

Description of the construction of this hyperbola in the plane, including the proof that the curve described this way is the hyperbola determined by the equation.

Examples of the reduction of equations to the forms of Theorems XII and XIII

1. Reduction of the equation

$$y^2 + \frac{2bxy}{a} + 2cy = \frac{fx^2}{a} + ex + d^2.$$

The equation is first reduced to

$$\frac{a^2 z^2}{fa + b^2} = v^2 - h^2 - \frac{a^2 d^2 + a^2 c^2}{fa + b^2} \quad ,$$

where $\qquad z = y - \dfrac{bx}{a} - c \; , \; v = x - \dfrac{a^2 e + 2abc}{2fa + 2b^2} \quad ,$

and $\qquad 2h = \dfrac{a^2 e + 2abc}{fa + b^2} \; .$

Then two cases are distinguished:

(i) $\qquad h^2 > \dfrac{a^2 d^2 + a^2 c^2}{fa + b^2} \; ;$

(ii) $\qquad h^2 < \dfrac{a^2 d^2 + a^2 c^2}{fa + b^2} \; .$

In the first case the curve turns out to be a hyperbola opened towards the x-axis. In the second case the curve turns out to be a hyperbola with the rounded side towards the x-axis.

A detailed *descriptio* with complete *demonstratio* is given for these curves.

2. Reduction of the equation

$$x^2 + 2ay = \frac{2bxy}{a}$$

to the form $\qquad z^2 - \dfrac{a^6}{b^4} = \dfrac{a^2 v^2}{b^2} \quad ,$

where $\qquad z = y - \dfrac{a^3}{b^2}$ and $v = x - \dfrac{by}{a} \, .$

Again this includes a detailed *determinatio* and *demonstratio*.

Problem II. *Proposition* 16.

Given two points, find a third point with the property that the line segments drawn from this point to each of the two given points differ by a given distance and determine and describe the locus to which the required point belongs.

This locus turns out to be a hyperbola, as can be concluded from the equation that is obtained, which is treated in Theorem XII. This can also be verified using the definition of a hyperbola derived from the "application problems" from Greek antiquity. The two given points are designated as foci.

Jan de Witt tacitly assumes that the solution lies in the plane.

Corollary 1. If from a point selected at random on a hyperbola, segments are drawn to both umbilici, the longest of them will exceed the shortest by the length of the transverse axis.

Corollary 2. If from a point selected at random on a hyperbola, straight lines are drawn to both umbilici, then the line that bisects the angle enclosed by these straight lines touches the curve at this point and conversely.

Example of the reduction of equations to the form of Theorem XIV

Reduction of the equation

$$y^2 + \frac{2bxy}{a} - 2cy = -x^2 + dx + k^2$$

to the form

$$z^2 = \frac{-a^2x^2 + b^2x^2}{a^2} + \frac{dax - 2bcx}{a} + c^2 + k^2,$$

where

$$z = y - c + \frac{bx}{a}.$$

This equation is then reduced to

$$\frac{a^2z^2}{a^2 - b^2} = -v^2 + h^2 + \frac{c^2a^2 + k^2a^2}{a^2 - b^2},$$

where

$$2h = \frac{da^2 - 2bca}{a^2 - b^2}$$

and

$$v = x - h$$

Assuming that $a^2 > b^2$, this equation is further reduced to

$$\frac{lz^2}{g} = f^2 - v^2,$$

where

$$\frac{l}{g} = \frac{a^2}{a^2 - b^2}$$

and

$$f^2 = h^2 + \frac{c^2a^2 + k^2a^2}{a^2 - b^2}.$$

The equation has now been reduced to the form of Theorem XIV and the conclusion is that it represents an ellipse or a circle. Again a detailed *descriptio* and the associated *demonstratio* are given.

The case $a^2 < b^2$ is not considered; indeed, it would not lead to an example of Theorem XIV.

Problem III. *Proposition* 17.

Given two points, find a third point with the property that the segments drawn from this point to each of the given points are, taken together, equal to a given length, and determine and describe the locus to which the required points belong.

This locus turns out to be an ellipse, as can be concluded from the equation that is obtained, which is treated in Theorem XIV. This can also be verified using the definition of an ellipse derived from the "application problems" from Greek antiquity. Again the two given points are designated as foci.

Jan de Witt again tacitly assumes that the solution lies in the plane.

Corollary 1. Taken together, the segments drawn from an arbitrary point on an ellipse to each of the umbilici are equal to the length of the transverse axis.

Corollary 2. If one draws straight lines from an arbitrary point on an ellipse to each of the umbilici, then the exterior bisection of the resulting angle will touch the curve at the aforementioned point.

Conversely, the tangent at this point forms equal angles with the extended line segments joining this point to the umbilici.

Chapter IV

General Rule to find and determine arbitrary plane and solid loci

In this chapter, a classification of the general equation of degree at most two in the undetermined quantities x and y is undertaken, and for each equation the corresponding line or curve is examined.

For the terms "plane" and "solid" loci, see the Introduction.

Jan de Witt distinguishes the following cases:

1.
$$y = \frac{bx}{a} \qquad \text{or } y = x, \text{ if } a = b.$$

$$y = \frac{bx}{a} \pm c \quad \text{or } y = c - \frac{bx}{a}.$$

He notes that one of the two quantities x or y may be missing. This remark is necessary because a and b are known *positive* quantities (line segments).

2.
$$\begin{array}{ll}
y^2 = dx & dy = x^2 \\
y^2 = dx \bullet f^2 & \quad\text{or conversely}\quad dy \bullet f^2 = x^2 \\
z^2 = dx & dy = v^2 \\
z^2 = dx \bullet f^2 & dx \bullet f^2 = v^2
\end{array}$$

$$y^2 = \frac{lx^2}{g} \bullet f^2 \qquad\qquad yx = f^2$$

$$z^2 = \frac{lx^2}{g} \bullet f^2 \qquad\qquad zx = f^2$$

3.   or even

$$y^2 = \frac{lv^2}{g} \bullet f^2 \qquad\qquad yv = f^2$$

$$z^2 = \frac{lv^2}{g} \bullet f^2 \qquad\qquad zv = f^2$$

Remarks:

1.   A term of the form $A \bullet B$ represents three cases, namely, $A + B$, $A - B$, and $-A + B$. The case $-A - B$ does not occur as A and B, as well as the left-hand side, must all be positive quantities.

2.   In addition to the original variables x and y, variables v and z also occur in these equations. Two cases must be distinguished:

   (i) z has the form $z = y \pm c$, $z = y \pm bx/a$, or $z = y \pm bx/a \pm c$, in which case v has the form $v = x \pm h$, and therefore does not contain any term with y.

   (ii) v has the form $v = x \pm c$, $v = x \pm v = x \pm by/a$, or $v = x - by/a \pm c$, in which case z has the form $z = y \pm h$, and therefore does not contain any term with x.

   Here a, b, c, and h are known positive quantities (line segments).

3.   The general Rule states that every quadratic equation in the unknown quantities x and y can be reduced to one of the forms mentioned in 2 and 3. For the forms $zx = f^2$ and $yv = f^2$ we only allow $z = y \pm h$ and $v = x \pm k$. De Witt does not prove this rule explicitly; rather, he refers to the methods applied in the examples. In what follows he shows that the equations mentioned in (2) and (3) all represent conics. See also the Introduction, Section 1.10.

In order to help the reader make their way through this chapter, let us first state its global layout

The following are treated in succession:

   (i)   from p. [306] to [307] – the straight line;

   (ii) from p. [308] to [314], line 4 from the bottom – the parabola given by one of the equations in the first column in 2, p. [305] (p. 41 of this summary);

   (iii) from p. [314], line 3 from the bottom, to p. [318], line 9 from the top – the parabola given by one of the equations in the second column in 2; that is, the converse of (11), where x and y have been interchanged;

(iv)   from p. [318], line 10 from the top, to p. [330] – the equations in the first column in 3 where the term with x^2 or v^2 has a plus sign; these equations turn out to represent hyperbolas;

(v)   from p. [331] to [332], line 7 from the bottom – the equations in the second column in 3, which also turn out to represent hyperbolas;

(vi)   from p. [332], line 6 from the bottom to the end of the chapter – the equations in the first column in 2 in which the term with x^2 or v^2 has a minus sign, while obviously f^2 has a plus sign; these equations turn out to represent ellipses or circles.

As in previous chapters Jan de Witt chooses an axis AB along which the abscissas x are measured out, in positive direction only, and an angle ABE under which the ordinates y are positioned. Of the curves treated here, only the part above the axis is considered.

The straight line

In case 1 on p. [305] (p. 42 in the summary) – the straight line – the following cases are distinguished and illustrated:

$$y = x, \quad y = \frac{bx}{a}, \quad y = \frac{bx}{a} + c, \quad y = \frac{bx}{a} - c, \quad y = c - \frac{bx}{a}.$$

First the constructions are given in the form of lists of instructions; this is followed by a meticulous proof that the abscissa and ordinate of any point on one of the constructed lines satisfy the equation in question. For an illustration, see Figure 2.3 (p. [307] of the text).

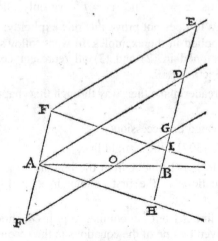

FIGURE 2.3

The parabola

In case 2 – the parabola – the left column is treated first. Nine cases are distinguished, each of which is split up into subcases. Jan de Witt begins with a list of strict instructions for the constructions. At the end of those he remarks that the proof that the constructed curves satisfy the corresponding equations is not difficult (*Quorum quidem omnium demonstratio perfacilis est*, p. [311], line 6 from the top). Only then does he begin with these proofs.

As far as the corresponding illustrations are concerned, note that Jan de Witt only gives the positions of the transverse axes with associated conjugate axes and vertices, but never draws a curve.

Moreover, for a given case he uses the same letter for the vertex in all associated subcases. For example, in Case IX (p. [310] to p. [314]) we can distinguish nine subcases; in each of them the letter Q denotes the vertex of the corresponding parabola.

Again the construction of the curves characterized above by equations is given without proof, in the form of lists of instructions.

This concerns the following situations: the figure that corresponds to Cases I to IX is Figure 2.4 (p. [308]).

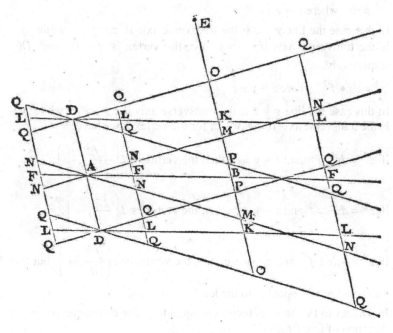

FIGURE 2.4

I. $y^2 = dx$

This equation represents a parabola with transverse axis AB and vertex A, while the ordinate-wise applied lines make an angle with this axis that is equal to the given or chosen angle ABE.

II. $y^2 = dx \bullet f^2$

This equation represents a parabola with transverse axis on AB. For the vertex F we have:

if $y^2 = dx + f^2$, then the vertex is $F\left(-\dfrac{f^2}{d},0\right)$;

if $y^2 = dx - f^2$, then the vertex is $F\left(\dfrac{f^2}{d},0\right)$;

if $y^2 = -dx + f^2$, then the vertex is $F\left(\dfrac{f^2}{d},0\right)$.

In the last case the parabola "has its opening to the left."

III. $z^2 = dx$, where $z = y \pm c$

In this case the line $y = c$ is the transverse axis if $z = y - c$, while $y = -c$ is the transverse axis if $z = y + c$; the vertex is $D(c,0)$ and $D(-c,0)$ respectively.

IV. $z^2 = dx \bullet f^2$, where $z = y \pm c$

In this case the line $y = c$ is the transverse axis if $z = y - c$, while $y = -c$ is the transverse axis if $z = y + c$; for the vertex L we have:

if $z^2 = dx + f^2$ and $z = y \pm c$, then the vertex is $L\left(-\dfrac{f^2}{d},\mp c\right)$;

if $z^2 = dx - f^2$ and $z = y \pm c$, then the vertex is $L\left(\dfrac{f^2}{d},\mp c\right)$;

if $z^2 = -dx + f^2$ and $z = y \pm c$, then the vertex is $L\left(\dfrac{f^2}{d},\mp c\right)$, but then the

parabola "has its opening to the left."

In Cases I to IV the latus rectum is equal to d; the conjugate axis lies in the direction of line BE.

V. $z^2 = dx$, where $z = y \pm (bx/a)$

In this case one chooses a point M on BE so that $AB : BM = a : b$. The point M lies "above" the line AB if $z = y - (bx/a)$ and "below" the line AB if $z = y + (bx/a)$; the support of AM is then the transverse axis, the conjugate axis lies in the direction of BE; the corresponding vertex is A.

VI. $z^2 = dx \bullet f^2$, where $z = y \pm (bx/a)$

Again one chooses AM as in Case V, but next one draws the lines FL through the points F and L (see II and IV for the definitions). These lines meet the lines AM at the points N. The following cases are distinguished with respect to the position of the parabola:

1. $z^2 = dx + f^2$ and $z = y + (bx/a)$ — the line AM with equation $y = -(bx/a)$ is the transverse axis, the corresponding vertex N has abscissa $AF = -f^2/d$

2. $z^2 = dx - f^2$ and $z = y + (bx/a)$ — the line AM with equation $y = -(bx/a)$ is the transverse axis, the corresponding vertex N has abscissa $AF = f^2/d$

3. $z^2 = -dx + f^2$ and $z = y + (bx/a)$ — the line AM with equation $y = -(bx/a)$ is the transverse axis, the corresponding vertex N has abscissa $AF = (f^2/d)$, but now the parabola "has its opening to the left"

The cases $z^2 = dx + f^2$, $z^2 = dx - f^2$, and $z^2 = -dx + f^2$, where $z = y - (bx/a)$, are treated analogously.

VII. $z^2 = dx$, where $z = y + bx/a + c$ or $z = y - bx/a - c$

1. $z^2 = dx$, $z = y + bx/a + c$ — the vertex is $D(-c, 0)$; the transverse axis $y = -bx/a - c$.

2. $z^2 = dx$, $z = y - bx/a - c$ — the vertex is $D(c, 0)$; the transverse axis $y = bx/a + c$

VIII. $z^2 = dx$, where $z = y + bx/a - c$ or $z = y - bx/a + c$

1. $z^2 = dx$, $z = y + bx/a - c$ — the vertex is $D(c, 0)$; the transverse axis $y = -bx/a + c$

2. $z^2 = dx$, $z = y - bx/a + c$ — the vertex is $D(-c, 0)$; the transverse axis $y = bx/a - c$

In Cases VII and VIII the conjugate axes lie parallel to BE.

IX. $z^2 = dx \bullet f^2$, where $z = y \pm (bx/a) \pm c$

1. $z^2 = dx + f^2$ – the abscissa of the vertex Q is $-f^2/d$

2. $z^2 = dx - f^2$ – the abscissa of the vertex Q is f^2/d

3. $z^2 = -dx + f^2$ – the abscissa of the vertex Q is f^2/d, but now the parabola "has its opening to the left."

The position of the corresponding axes and vertices can be found as described in VII and VIII. In all this concerns nine subcases.

Finally we find the statement that in cases V to IX the latus rectum p satisfies the proportion $e : a = d : p$, where e is defined by $AB : BM : AM = a : b : e$ (see Figure 2.4).

In each of these nine cases the parabolas are described through lists of instructions that use the parameters of the given equation. After the statements above, Jan de Witt proves that these parabolas, which seem to appear out of nowhere, are indeed represented by the equations he started out with.

Jan de Witt concludes his treatment of the parabola with a short discussion of the "converse" cases, where x and y have been interchanged; that is:

$$dy = x^2, \; dy \bullet f^2 = x^2, \; dy = v^2, \; dy \bullet f^2 = v^2,$$

where successively

$$v = x \pm c, \; v = x \pm \frac{by}{a}, \text{ or } v = x \pm \frac{by}{a} \pm c.$$

The hyperbola

In case 3 on p. [305] (p. 41 of this summary) Jan de Witt first considers the equations in the first column that represent a hyperbola, that is, whose terms with x^2 or v^2 have a plus sign. Then he considers the equations in the second column that also represent a hyperbola; that is:

$$yx = f^2, \; zx = f^2, \; yv = f^2 \text{ and } zv = f^2,$$

where

$$z = y \pm c \text{ and } v = x \pm h.$$

A. From p. [318] on the following modifications of the first column in 3, p. [305], are distinguished.

I. $lx^2/g = y^2 - f^2$ or $ly^2/g = x^2 - f^2$

II. $lx^2/g = z^2 - f^2$ or $lz^2/g = x^2 - f^2$, where

1. $z = y \pm c$

2. $z = y \pm (bx/a)$

3. $z = y \pm (bx/a) \pm c$

III. $lv^2/g = y^2 - f^2$ or $ly^2/g = v^2 - f^2$, where $v = x \pm h$

IV. $lv^2/g = z^2 - f^2$ or $lz^2/g = v^2 - f^2$, where $v = x \pm h$ and

1. $z = y \pm c$

2. $z = y \pm (bx/a)$

3. $z = y \pm (bx/a) \pm c$

As one can see the cases that are distinguished are those where x and y, or x and z, y and v, or z and v have been interchanged. Again one receives the description of such a hyperbola as a list of instructions, followed by the proof that the coordinates of the points on it satisfy the initial equation.

Here are the results (see also Figure 2.6).

I. $lx^2/g = y^2 - f^2$ or $ly^2/g = x^2 - f^2$

In the first case one chooses the transverse axis of the hyperbola along the line AX, through A and parallel to the line BE; in the second case one chooses the transverse axis along the line AB. The associated conjugate axes are parallel to respectively AB and BE. In both cases A is the center and $2f$ the length of the transverse diameter.

The associated latus rectum p is determined (that is, chosen) using the proportion $2f : p = l : g$. The length $2d$ of the conjugate diameter then follows from the definition of the latus rectum using the proportion $2f : 2d = 2d : p$.

In our further explanation and proof of the correctness of the construction we restrict ourselves to the second equation that was mentioned; Jan de Witt of course treats both equations. The point B on the curve is chosen as the intersection point of the curve with the line through B that makes the given angle with the abscissa-axis AE. Jan de Witt now refers to the characteristic property of the hyperbola that he mentioned in *Liber Primus* as Theorem IX, Proposition 10, p. [196], illustrated with the figure on p. [198] (here Figure 2.5). For the situation in Figure 2.5, this characteristic property implies that:

$$ND^2 : NC \cdot NP = GH^2 : CP^2 = CH^2 : AC^2, \text{ that is :}$$

$$ND^2 : (NA - NC)(NA + NC) = CH^2 : AC^2$$

If we now set $ND = y$, $NA = x$, $AC = f$, $CH = d$, then this property implies that

$$y^2 : (x - f)(x + f) = d^2 : f^2.$$

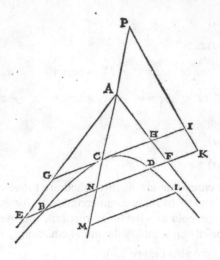

FIGURE 2.5

The ratio $d^2 : f^2$ can be determined as follows. By the choice of p we have $2f : p = l : g$ and by the definition of p we have

$$2f : 2d = 2d : p,$$

so that $d^2 : f^2 = p : 2f = g : l$

and therefore

$$\frac{ly^2}{g} = x^2 - f^2 .$$

In passing let us remark that Jan de Witt treats the cases $l = g$ and $l \neq g$ separately, just as further on, in his treatment of the ellipse, where he treats the circle as a distinct case.

Again he only shows that the coordinates of the points on the constructed curve satisfy the equation, but does not search for all points whose coordinates have this property. As a consequence he only finds one branch of the hyperbola.

II.1. $lx^2 / g = z^2 - f^2$ or $lz^2 / g = x^2 - f^2$, where $z = y \pm c$

In this case one chooses $D(0, c)$ as center if $z = y - c$ and $D(0, -c)$ if $z = y + c$. After this the reasoning is analogous to that in I, where D takes the place of A. This now concerns hyperbolas with transverse axis $x = 0$ and $y = c$ if $z = y - c$ and hyperbolas with transverse axis $x = 0$ and $y = -c$ if

$z = y + c$. The curves have been translated over a distance of $\pm c$ in the ordinate direction.

II.2. $lx^2 / g = z^2 - f^2$ or $lz^2 / g = x^2 - f^2$, where $z = y \pm (bx/a)$

In both cases one chooses the point A as center. Then one chooses the line AM with equation $y = bx/a$ if $z = y - (bx/a)$, and the line AM with equation $y = -bx/a$ if $z = y + (bx/a)$ (see Figure 2.7, which is the figure on p. [323]).

In the first case the transverse axis of the required hyperbola lies on the line AW, parallel to BE. The conjugate axis lies in the direction of AM (corresponding to $z = y + (bx/a)$ or $z = y - (bx/a)$).

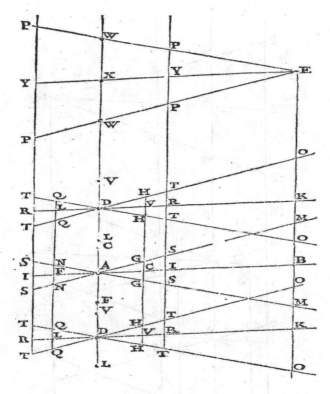

FIGURE 2.6

In the second case the transverse axis of the required hyperbola lies on the line AM (corresponding to $z = y + (bx/a)$ or $z = y - (bx/a)$). The conjugate axis lies in the direction of AW.

In the first case the length of the transverse diameter of the hyperbola (on AW) is equal to $2f$. The corresponding latus rectum p is determined by the proportion

$$2f : p = a^2 l : e^2 g, \text{ where we have}$$

$$AB : BM : AM = a : b : e.$$

The length $2d$ of the conjugate diameter follows from the definition of the latus rectum as the third element of the proportion involving $2f$ and $2d$, that is, from

$$2f : 2d = 2d : p.$$

The ratio $d^2 : f^2$ that is so important is then equal to $e^2 g / a^2 l$.

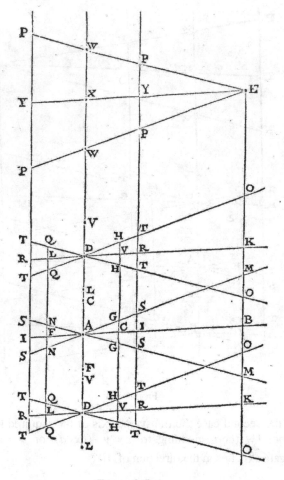

FIGURE 2.7

In the second case the length $2m$ of the transverse diameter of the hyperbola (on AM) is equal to $2ef/a$. The vertices therefore lie at the intersection points of the lines AM with the lines $x = \pm a$. In this case the corresponding latus rectum p is determined by the proportion $2m : p = e^2l : a^2g$. The length $2d$ of the conjugate diameter again follows from the definition of the latus rectum as the third element of the proportion involving $2m$ and $2d$, that is, from $2m : 2d = 2d : p$. The ratio $d^2 : m^2$ is then equal to a^2g/e^2l.

This description of the required hyperbola is again followed by a proof that the coordinates of the points on the constructed curve indeed satisfy the initial equations.

II.3. $lx^2/g = z^2 - f^2$ or $lz^2/g = x^2 - f^2$, where $z = y \pm (bx/a) \pm c$.

The reasoning is analogous to that in II.2, where $D\,(0,\,c)$ takes the place of A if $z = y \pm (bx/a) - c$ and $D(0,\,-c)$ if $z = y \pm (bx/a) + c$. Here too the curves have been translated over a distance of $\pm c$ in the ordinate direction.

III. $lv^2/g = y^2 - f^2$ or $ly^2/g = v^2 - f^2$, where $v = x \pm h$

In this case one chooses $I\,(h,\,0)$ as center if $v = x - h$ and $I\,(-h,\,0)$ if $v = x + h$. The reasoning is analogous to that in I, where it now concerns hyperbolas with transverse axis on $x = h$ and $y = 0$ if $v = x - h$, and hyperbolas with transverse axis on $x = -h$ and $y = 0$ if $v = x + h$. The curves have been translated over a distance of $\pm h$ in the abscissa direction.

IV.1. $lv^2/g = z^2 - f^2$ or $lz^2/g = v^2 - f^2$,, where $v = x \pm h$ and $z = y \pm c$.

In this case one chooses $R(\,\pm h,\,\pm c\,)$ as center, where the signs correspond to those chosen in

$$v = x \pm h \text{ and } z = y \pm c$$

The reasoning is analogous to that in I; this again concerns hyperbolas to which a parallel translation has been applied.

IV.2. $lv^2/g = z^2 - f^2$ or $lz^2/g = v^2 - f^2$, where $v = x \pm h$ and $z = y \pm (bx/a)$

In this case one chooses $S(\pm h, \pm (bh/a))$ as center, where the signs correspond to those chosen in $v = x \mp h$ and $z = y \mp (bx/a)$. The reasoning is analogous to that in II.2.

IV.3. $lv^2/g = z^2 - f^2$ or $lz^2/g = v^2 - f^2$, where

$$v = x \pm h \text{ and } z = y \pm (bx/a) \pm c$$

In this case one chooses $T(\pm h, \pm (bh/a) \pm c)$ as center, where the signs correspond to those chosen in $v = \mp h$ and $z = y \mp (bx/a) \mp c$. The reasoning is analogous to that in II.3.

B. Four other cases where the required locus is a hyperbola

This concerns the equations

1. $yx = f^2$;

2. $zx = f^2$;

3. $vv = f^2$;

4. $zv = f^2$,

where $z = y \pm h$ and $v = x \pm c$ and f, h, and c are known positive quantities (line segments).

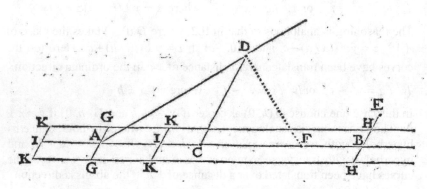

FIGURE 2.8

1. $yx = f^2$

For the construction of the required hyperbola the line segment $AC = f$ is measured out on the abscissa-axis AB (see Figure 2.8). Then the line segment $CD = f$ is set out in the ordinate direction.

The required locus will turn out to be a hyperbola with center A, with transverse axis, of length $2AD$, lying along the support of AD (hence with vertex D) and with asymptote AB.

That the coordinates of the points on this curve satisfy the equation $yx = f^2$ simply follows from the characteristic property of the hyperbola that one finds in *Liber Primus*, Theorem 3, on p. [181]. Indeed, if we let E be the intersection point of the curve with the line through B under the given angle (in the ordinate direction), then by the property mentioned above $AB \cdot BE = AC^2$, that is, $xy = f^2$.

2. $zx = f^2$, where $z = y \pm h$

In this case one does not choose A as vertex but the point $G(0, \pm h)$, where the sign corresponds to that chosen in $z = y \mp h$. The reasoning is analogous to that in (1).

3. $vv = f^2$, where $v = x \pm c$

In this case one does not choose A as vertex but the point $I(\pm c, 0)$, where the sign corresponds to that chosen in $v = x \mp c$. The reasoning is analogous to that in (1).

4. $zv = f^2$, where $z = y \pm h$ and $v = x \pm c$

In this case one does not choose A as vertex but the point $K(\pm c, \pm h)$, where the signs correspond to those chosen in $v = x \mp c$ and $z = y \mp h$. Again the reasoning is analogous to that in (1).

The ellipse

As mentioned before, this chapter and consequently the whole work, concludes with a discussion of the remaining cases in the first column in 3 on p. [305] (pp. 40–41 of this summary); that is: those where the term with x^2 or that with v^2 has a minus sign. The term f^2 then of course has a plus sign. These equations turn out to represent ellipses or circles.

Jan de Witt does not take exactly the equations in the column in question, but starts with the formula $ly^2 / g = f^2 - x^2$ and treats it together with a number of modifications. He distinguishes the following cases:

I. $ly^2 / g = f^2 - x^2$

II. $lz^2 / g = f^2 - x^2$, where

    1. $z = y \pm c$

    2. $z = y \pm (bx / a)$

    3. $z = y \pm c \pm (bx / a)$

III. $ly^2 / g = f^2 - v^2$, where $v = x \pm h$

IV. $lz^2 / g = f^2 - v^2$, where $v = x \pm h$ and

    1. $z = y \pm c$

    2. $z = y \pm (bx / a)$

    3. $z = y \pm c \pm (bx / a)$

Here are the results he obtains (see Figure 2.9).

I. $ly^2 / g = f^2 - x^2$

The claim is that this concerns an ellipse that can be described as follows: The center is A, the transverse diameter FAC lies on the support of AB and is of length $2f$, so that $FA = AC = f$. The conjugate diameter lies in the direction of line BE, and the corresponding latus rectum is determined by the condition $2f : p = l : g$.

To prove that the described curve is the required locus, Jan de Witt takes a point (x,y) on the curve and using his geometric definition of an ellipse (see *Liber Primus*, Theorem XII, p. [205]) shows that the coordinates of this point satisfy the equation in question. He also remarks that if $l = g$ and the angle ABE is a right angle, then the curve is a circle.

We note that once more he has not proved that he has found all points that satisfy the equation.

II.1. $lz^2 / g = f^2 - x^2$, where $z = y \pm c$

This concerns an ellipse (or a circle) with center $D(0, c)$ if $z = y - c$ and $D(0, - c)$ if $z = y + c$.

The reasoning is analogous to that in I, where D takes the place of A.

II.2. $lz^2 / g = f^2 - x^2$, where $z = y \pm (bx / a)$

Jan de Witt now constructs an ellipse as follows: The center is A, the transverse axis NAG lies on the support of AM, which has the following equation:

$$y = \frac{bx}{a} \text{ (if } z = y - \frac{bx}{a})$$

and

$$y = -\frac{bx}{a} \text{ (if } z = y + \frac{bx}{a}).$$

The conjugate axis lies in the direction of line BE. The length $2m$ of the transverse diameter NG is $2(ef / a)$, where we again have

$AB : BM : AM = a : b : e$ (see Figure 2.9).

The corresponding latus rectum is determined by the condition

$$2m : p = e^2 l : a^2 g.$$

To prove that the described curve is the required locus, Jan de Witt takes an arbitrary point (x, y) on the curve and using his geometric definition of an ellipse shows that the coordinates of this point satisfy the equation in question.

He also remarks that if $e^2 l = a^2 g$ and the angle AME is a right angle, then the curve is a circle.

II.3. $lz^2 / g = f^2 - x^2$, where $z = y \pm c \pm (bx / a)$

This concerns an ellipse (or a circle) with center D $(0, c)$ if $z = y - c \pm (bx/a)$ and D $(0, -c)$ if $z = y + c \pm (bx/a)$. The reasoning is analogous to that in II.2, where D takes the place of A.

III. $ly^2/g = f^2 - v^2$, where $v = x \pm h$

This concerns an ellipse with center $I(h, 0)$ if $v = x - h$ and $I(-h, 0)$ if $v = x + h$. The reasoning is analogous to that in I, where I takes the place of A.

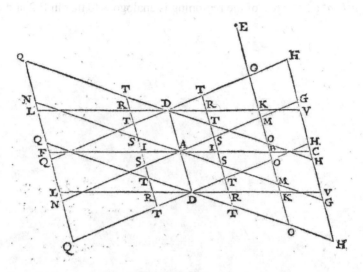

FIGURE 2.9

IV.1. $lz^2/g = f^2 - v^2$, where $v = x \pm h$ and $z = y \pm c$

This concerns an ellipse with center $R(\pm h, \pm c)$, where the signs correspond to those chosen $v = x \mp h$ and $z = y \mp c$. The transverse diameter lies on the line R parallel to AB and has length $2f$. The corresponding latus rectum is determined by the condition that the ratio of the transverse diameter to this latus rectum is as $l : g$. The reasoning is analogous to that in II.1, where one of the four points R takes the place of A.

IV.2. $lz^2/g = f^2 - v^2$, where $v = x \pm h$ and $z = y \pm (bx/a)$.

This concerns an ellipse whose center S is defined as the intersection point of the line $x = \pm h$ and the line $y = \pm bx/a$. The transverse diameter lies on the line $y = \pm bx/a$ and has length $2ef/a$. The corresponding latus rectum is determined by the condition that the ratio of the transverse diameter to this latus rectum is as $e^2 l : a^2 g$. The reasoning is analogous to that in II.2.

IV.3. $lz^2 / g = f^2 - v^2$, where $v = x \pm h$ and $z = y \pm c \pm (bx / a)$.

This concerns an ellipse whose center T is defined as the intersection point of the line $x = \pm h$ and the line through D ($\pm h$, 0) with equation $y = \pm h \pm bx / a$. The transverse diameter lies on the line $y = \pm h \pm bx / a$ and has length $2ef / a$. The corresponding latus rectum is determined by the condition that the ratio of the transverse diameter to this latus rectum is as $e^2 l : a^2 g$. The rest of the reasoning is analogous to that in II.2 and II.3.

3

Latin text and translation

A.W. Grootendorst et al. (eds.), *Jan de Witt's Elementa Curvarum Linearum*,
Sources and Studies in the History of Mathematics and Physical Sciences,
DOI 10.1007/978-0-85729-142-4_3, © Springer-Verlag London Limited 2010

[243]

JAN DE WITT'S

FOUNDATIONS OF CURVED LINES

SECOND BOOK

CHAPTER I

GENERAL PROPOSITION

I N every problem in which it is asked to investigate a locus, be this on a straight line or on a curve, one arrives at an equation determining an arbitrarily chosen point on the required locus [1.1] if one considers as known and determinate two unknown and indeterminate line segments enclosing a given or chosen angle [1.2].

If in this equation (after reduction to its simplest form) neither of the two unknown quantities has been raised to the second or a higher power − that is, if they do not occur multiplied by themselves nor by each other − then the required locus will be a straight line. But if one of these unknown quantities has been raised to the second power but the other not and if this other one has not been multiplied by itself nor by the first one, then the required locus will be a parabola.

If, however, each of the two unknown quantities has been raised to the second power or if a mutual product occurs in the equation (in fact, the equation is not of a higher degree if it concerns a plane or solid locus [1.3]), then the required locus will be a hyperbola or an ellipse or the circumference of a circle.

IOHANNIS DE WITT
ELEMENTA
CVRVARVM
LINEARVM.

LIBER SECVNDVS.

CAPVT I.
PROPOSITIO GENERALIS.

IN omni quæstione, ubi indagandus proponitur Locus, five is fit ad lineam rectam, five ad curvam, fuppofitis duabus lineis rectis incognitis atque indeterminatis, datum vel affumptum angulum comprehendentibus, tanquam cognitis ac determinatis, devenitur ad Æquationem, affumptum quodlibet quæfiti Loci punctum determinantem; in qua quidem æquatione, poftquam ad fimpliciffimos terminos erit reducta, fi neutra incognitarum ad duas plurefve dimenfiones affurgat, hoc eft, fi neque in fe, neque in alteram incognitam ducta feu multiplicata reperiatur, quæfitus Locus erit linea recta: At fi earundem incognitarum altera ad quadratum afcendat, altera verò non item, fed neque in fe, neque in alteram incognitam ducta fit, erit Locus quæfitus Parabola. Quòd fi verò utraque ad quadratum afcendat, five altera in alteram ducta in æquatione reperiatur (altiùs enim æquatio non affurget, fi de loco Plano Solidovè quæftio fit): erit Locus quæfitus vel Hyperbola, vel Ellipfis, vel Circuli circumferentia.

Hh 2   Quo-

[244]

Of all these curves, however, a separate determination, definition, and demonstration can be given in various ways [1.4], but it will be sufficient to give one of the most simple and most general of them.

In the first case, in fact, when neither of the unknown quantities has been raised to the second or a higher power, the equation can be reduced to one of the following forms in which one of the unknown quantities is expressed by x, the other by y [1.5]:

I. $y = \dfrac{bx}{a}$, or (if we put $a = b$) $y = x$

II. $y = \dfrac{bx}{a} + c$, or, with the same supposition as before, $y = x + c$

III. $y = \dfrac{bx}{a} - c$, or $y = x - c$

IV. $y = -\dfrac{bx}{a} + c$, or $y = -x + c$

However, as usual, let the following rule be assumed with respect to the unknown quantities: the initial point of one of them, let us say of x, will be fixed and immutable and this quantity is understood to extend indefinitely from that fixed and immutable point along a straight line the position of which is given. The other quantity, the length of which is indeterminate as well, is connected to the former line at its unfixed endpoint at a given or chosen angle [1.6]. This being supposed, it seems possible to propose, to determine, and to demonstrate the aforesaid in a suitable way by means of the following theorems.

Theorem I

Propositon 1

If the equation is $y = bx/a$, then the required locus will be a straight line.

Let, in fact, A be the immutable initial point of x [1.7] and let it be understood that x extends indefinitely along the line AB. Then we select at random a point on this AB, let us say B, and we draw BC

244   ELEM. CVRVARVM

Quorum quidem omnium particularis determinatio,
descriptio, & demonstratio variis modis fieri potest;
at verò ex simplicissimis, generalissimisque aliquem
annotasse suffecerit.

Ac primò quidem casu, cùm neutra quantitatum
incognitarum ad duas pluresve dimensiones ascendit, si
earum una exprimatur per x, atque altera per y, potest
æquatio ad aliquam sequentium formularum reduci.

I. $y \infty \frac{bx}{a}$, sive (posito $a \infty b$) $y \infty x$.

II. $y \infty \frac{bx}{a} + c$, sive, posito, ut supra, $y \infty x + c$.

III. $y \infty \frac{bx}{a} - c$, sive $y \infty x - c$.

IV. $y \infty - \frac{bx}{a} + c$, sive $y \infty - x + c$.

Fiat autem earundem quantitatum incognitarum secundùm regulam talis assumptio, ut initium unius,
verbi gratiâ, ipsius x, certum sit & immutabile, utque
eadem illa quantitas ex certo & immutabili illo initio
in linea recta positione data intelligatur indefinitè extendi, altera verò indeterminatæ quoque longitudinis
linea priori in extremitate incerta in dato vel assumpto angulo conjungi. Quibus quidem suppositis, ea,
quæ prædicta sunt, sequentibus Theorematis non incongruè proponi, determinari, ac demonstrari posse
videntur.

THEOREMA I.

Propositio 1.

Si æquatio sit $y \infty \frac{bx}{a}$, erit locus quæsitus linea recta.

Sit enim ipsius x initium immutabile punctum A, atque eadem
illa x per rectam A B indefinitè se extendere intelligatur. Dein,
sumpto in eadem A B puncto utcunque, veluti B, agatur B C in
angulo

[245]

at the angle *ABC*, equal to the given or chosen angle such that the ratio of the intercepted *AB* to the drawn *BC* is the same as that of a known *a* to a known *b*; this means that *AB* is to *BC* as *a* is to *b*. Next, let line *AC* be drawn through *A* and *C*, infinitely produced [1.8], then this line will be the required locus. In fact, if we select a point at random on *AC*, let us say *D*, and if we draw *DE* at the angle *DEA* equal to the given or chosen angle and if we call this *DE* y, then[1] *AE* will be to *ED* as *AB* is to *BC*, that is as *a* to *b* and so is the ratio of x to y. Thus[2] $ay = bx$, that is, after dividing by a, $y = bx / a$ [1.9]. As point *D* has been selected at random on the line *AC*, the same proof will hold for all other points on the line *AC* and so *AC* is the required locus. And so not only the truth of the proposed theorem has been demonstrated, but the locus has also been determined [1.10].

THEOREM II

Proposition 2

If the equation is $y = (bx / a) + c$, then the required locus will be a straight line.

Let the assumptions and the constructions be the same as before and moreover, let the segment *AF* be drawn from *A* and parallel to *BC*, at the same side (of *AE*, transl.) as *BC* and equal to a known *c*. Let from *F* also be drawn *FG* parallel to *AC*. Then, I say, this *FG* is the required locus.

In fact, if we select a point at random on *FG*, let us say *G*, and if we draw *GE* at the angle *AEG*

---

[1] I, 29 and VI, 4. [2] VI, 16.

angulo A B C, ipſi dato vel aſſumpto æquali; ita ut eadem ſit ratio interceptæ A B ad ductam B C, quæ eſt *a* cognitæ ad *b* cognitam.

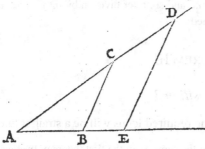

hoc eſt, ut ſit uti *a* ad *b*, ita A B ad B C. Denique per puncta A & C ducatur recta A C, indefinitè extenſa, eritque hæc ipſa locus quæſitus.

Etenim aſſumpto in A C puncto utcunque, veluti D, ductâque D E in angulo D E A, dato vel aſſumpto æquali, ſi eadem D E vocetur *y*, erit [1] ut A B ad B C, hoc eſt, ut *a* ad *b*, ita A E ad E D, hoc eſt, ita *x* ad *y*. Et fit [2] $a\,y \propto b\,x$, hoc eſt, dividendo utrinque per *a*, erit $y \propto \dfrac{b\,x}{a}$.

[1] *per 29 primi, & 4 ſexti.*

[2] *per 16 ſexti.*

Quare cum punctum D utcunque ſumptum ſit in linea A C, erit eadem de omnibus aliis lineæ A C punctis demonſtratio, ac proinde ipſa A C locus eſt quæſitus. Atque ita non ſolùm Theorematis propoſiti veritas demonſtrata, ſed & Locus quæſitus determinatus eſt.

T H E O R E M A II.
Propoſitio 2.

Si æquatio ſit *y* $\propto \dfrac{b\,x}{a} + c$, erit Locus quæſitus lineæ recta.

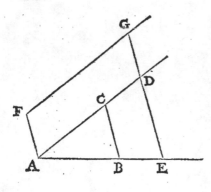

Poſitis, factisque, ut ſuprà, agatur inſuper ex A recta A F ipſi B C parallela, atque ad eaſdem cum ea partes, quæ ſit æqualis *c* cognitæ. Et ex F ductâ F G parallelâ A C, dico eandem F G eſſe Locum quæſitum.

Sumpto enim in F G puncto utcunque, veluti G, ductâque GE in angulo AEG, dato

Hh 3

[246]

(equal to the given or chosen angle), intersecting the straight line AC at D; and if this GE is called y, then $ED = y - c$.

In fact, as before[1] AE is to ED as AB is to BC, which means that x is to $y - c$ as a is to b and, therefore,[2] $ay - ac = bx$, or $ay = bx + ac$ and so, after division by a, $y = (bx / a) + c$ Which was to be demonstrated and determined.

THEOREM III

Proposition 3

If the equation is $y = (bx / a) - c$, then the required locus will be a straight line.

Let the assumptions and the constructions be the same as in the first theorem. In A we erect the segment AF, which is parallel to BC and is on the opposite side (of AB, transl.) from BC and which is equal to the known c. If then again FG is drawn from F parallel to AC, intersecting the straight line AB at H, then, I say, HG is the required locus.

Indeed, if we select a point at random on it, let us say G, and if we draw GE at the angle AEG (equal to the given or chosen angle), which being produced intersects AC at D and if we call GE y, then $ED = y + c$ [1.11]. By construction,[3] however, AE is to ED as AB is to BC, which means that x is to $y + c$ as a is to b. Therefore,[4] $ay + ac = bx$ or $ay = bx - ac$ and so, after division by a, $y = (bx / a) - c$. Which was proposed.

---

[1] VI, 29 and VI, 4. [2] VI, 16 [3] I, 29 and VI, 4 [4] VI, 16

246   E L E M. C V R V A R V M

dato vel aſſumpto æquali, quæ ſecet rectam A C in D, ſi eadem
G E vocetur y, erit E D ∞ y — c. At verò eſt, ut ſupra [1], uti A B
ad B C, ita A E ad E D, hoc eſt, ut a ad b, ita x ad y — c: ac propterea [2] ay — ac ∞ bx, vel ay ∞ bx + ac, adeoque, factâ diviſione per a, y ∞ $\frac{bx}{a}$ + c. Quod demonſtrandum determinandumque erat.

[1] per 29
primi, &
4 ſexti.
[2] per 16
ſexti.

T H E O R E M A   III.

Propoſitio 3.

Si æquatio ſit y ∞ $\frac{bx}{a}$ — c, erit Locus quæſitus linea
recta.

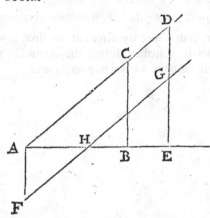

Poſitis factiſque ut in
Theoremate 1mo, agatur
inſuper ex A recta A F,
ipſi B C parallela, atque
ad oppoſitas cum ea par-
tes, quæ ſit æqualis c cognitæ. Et ex F ductâ ite-
rum F G ipſi A C paralle-
lâ, ſecante rectam A B in
H, dico H G eſſe Locum
quæſitum.

Sumpto enim in eadem
puncto utcunque veluti
G, ductâque G E in an-
gulo A E G, dato vel aſſumpto æquali, quæ producta ſecet A C
in D, ſi eadem G E vocetur y, erit E D ∞ a + c. Iam verò eſt [3]
ex conſtructione, ut A B ad B C, ita A E ad E D, hoc eſt, ut a ad
b, ita x ad y + c: ac propterea [4] ay + ac ∞ bx, vel a y ∞ bx — ac,
adeoque, factâ diviſione per a, y ∞ $\frac{bx}{a}$ — c. Quod eſt propoſitum.

[3] per 29
primi, &
4 ſexti.
[4] per 16
ſexti.

T H E O-

[247]

THEOREM IV

Proposition 4

If the equation is $y = c - (bx / a)$, then the required locus will be a straight line.

Let the assumptions and the constructions be the same as in the second theorem, with this difference that point C falls on the opposite side of AB and that the angle ABC is equal to the supplement of the given or chosen angle as is clear from the added figure; let the straight line FG be drawn from F parallel to AC, intersecting the straight line AB at H. Then, I say, FH is the required locus.

In fact, if we select a point at random on FH, let us say G, and if we draw GE at the angle AEG, equal to the given or chosen angle, which being produced intersects AC at D and if we call GE y, then $ED = c - y$. As by construction[1] AE is to ED, as AB is to BC, which means that x is to $c - y$ as a is to b, it follows,[2] that $ac - ay = bx$, or $ay = ac - bx$; that is, after division on both sides by a, $y = c - (bx / a)$. Which was proposed.

It can also happen, however, that before we arrive at the final equation one of the unknown quantities completely vanishes during the operation and that the other one remains as the only left, equal to a known quantity and

---

[1] I, 13 and 29 and VI, 4.   [2] VI, 16.

L i b. II. C a p. I. 247

T h e o r e m a I V.

Propofitio 4.

Si æquatio fit $y \infty c - \frac{bx}{a}$, erit Locus quæfitus linea recta.

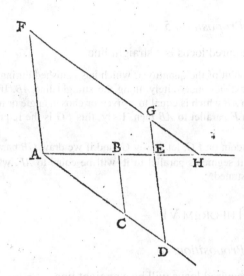

Pofitis, factisque, ut in Theoremate 2do, excepto quòd punctum C ab oppofita parte ipfius AB cadat, quodque angulus A B C æqualis fit dati vel affumpti anguli ad binos rectos complemento, quemadmodum in adjuncta figura apparet, agatur ex F recta F G ipfi A C parallela, occurrens rectæ A B in H : dico F H effe Locum quæfitum.

Sumpto enim in F H puncto utcunque, veluti G, ductâque G E in angulo A E G, dato vel affumpto æquali, quæ producta fecet A C in D, fi eadem G E vocetur y, erit E D ∞ $c - y$. Cumque fit ¹ ex conftructione, ut A B ad B C, ita A E ad E D, hoc eft, ut a ad b, ita x ad $c - y$: erit propterea ² $ac - ay \infty bx$, vel $ay \infty ac - bx$, id eft, dividendo utrinque per a, $y \infty c - \frac{bx}{a}$. Quod erat propofitum.

¹ *per 13 primi, &* 29 *primi, &*
² *ver 16 fexti.*
² *ver 16 fexti.*

At verò fieri etiam poteft , ut per operationem, priufquam ad æquationem deveniatur , quantitatum incognitarum altera penitus evanefcat , alteraque fola alicui cognitæ quantitati æqualis remaneat ; atque

exin-

[248]

that hence two formulae arise, which have to be reduced to the following forms, namely

1        $y = c,$

or

2. $x = c.$

THEOREM V

Proposition 5

If the equation is $y = c$, then the required locus is a straight line.

Let point A be the immutable initial point of the quantity x, which has vanished during the operation and let us suppose that x extends indefinitely along the straight line AB. If we next draw $AF=c$ from A at an angle to AB which is equal to a given or chosen angle or to its supplement and if we draw FG from F parallel to AB then, I say, this FG is the required locus.

Indeed, if we select a point at random on FG, let us say G, and if we draw GB parallel to AF, then it is clear that GB and all segments parallel to it[1] will be equal to AF, which means $y = c$. Which was to be demonstrated.

THEOREM VI

Proposition 6

If the equation is $x = c$, then the required locus will be a straight line.

Along the line AB, which is as before conceived as x, the segment AB of length equal to c is measured from the point A and from B the straight line BC is drawn, at a given or chosen angle. Then, I say, this BC, being produced indefinitely, will be the required locus.

Indeed, if we select a point

---

[1] I, 34

248   ELEM. CVRVARVM

exinde binæ infuper formulæ nafcuntur, quæ huc referri debent: nimirum,

  1. *y* ∞ *c*, vel
  2. *x* ∞ *c*.

THEOREMA V.
Propofitio 5.

Si æquatio fit *y* ∞ *c*, Locus quæfitus eft linea recta.

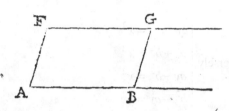

Sit quantitatis *x*, quæ per operationem evanuit, initium immutabile punctum A, atque eadem illa *x* per rectam A B indefinitè fe extendere intelligatur. Deinde ex A ductâ A F ∞ *c*, faciente cum A B angulum, ipfi dato vel affumpto aut ejufdem ad binos rectos fupplemento æqualem, fi ex F agatur F G ipfi A B parallela, dico eandem F G effe Locum quæfitum.

Etenim affumpto in F G puncto utcunque, veluti G, ductâque G B ipfi A F parallelâ, apparet eandem G B omnesque ipfi æquidiftantes [1] rectæ A F fore æquales, hoc eft, effe *y* ∞ *c*. Quod erat demonftrandum.

[1] *per* 34 *primi.*

THEOREMA VI.
Propofitio 6.

Si æquatio fit *x* ∞ *c*, erit Locus quæfitus linea recta.

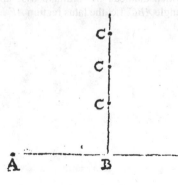

In linea A B, quæ, ut fupra, pro *x* concepta fit, fumatur à puncto A longitudo A B æqualis *c* cognitæ, atque ex B in dato vel affumpto angulo ducatur recta B C. dico eandem B C, indefinitè productam, effe Locum quæfitum.

Sumpto enim in eadem puncto

[249]

at random on it, let us say *C*, then by hypothesis *CB* will enclose an angle with the aforementioned *AB*, which is equal to the given or chosen angle, and so this *CB* can be called *y*. But by construction *CB* is and will always remain *AB*, which means $x = c$. Which was proposed.

CHAPTER II

F URTHER, in the second case (formulated above). where (as we know) in the equation (reduced to its simplest form) one of the unknown quantities has been squared, but the other not and the latter quantity is found neither multiplied by itself nor by the former, the equation can be reduced to one of the following forms:

$$
\left.
\begin{array}{ll}
\text{I.} & y^2 = ax \\
\text{II.} & y^2 = ax + b^2 \\
\text{III.} & y^2 = ax - b^2 \\
\text{IV.} & y^2 = -ax + b^2
\end{array}
\right\} \text{ or conversely}
\left\{
\begin{array}{l}
ay = x^2 \\
ay + b^2 = x^2 \\
ay - b^2 = x^2 \\
b^2 - ay = x^2
\end{array}
\right.
$$

Here it is supposed that *y* and *x* are the unknown quantities either conceived from the beginning or introduced later on, as we will explain at greater length shortly [2.1].

THEOREM VII

Proposition 7

If the equation is $y^2 = ax$ or conversely $ay = x^2$, then the required locus will be a parabola.

Let *A* be the immutable initial point of *x* and suppose that *x* extends indefinitely along the straight line *AB*. Let the given or chosen angle be equal to the angle *ABC*. Let firstly *AB* be chosen as a diameter of a parabola with which diameter the ordinate-wise applied lines enclose angles equal to the given or chosen angle *ABC*. Let the latus rectum *AF*

&to utcunque, velutiC, erit ex hypothesi CB cum priore A B comprehendens angulum A B C dato vel assumpto æqualem, poteritque proinde eadem C B vocari *y*. At verò est ex constructione, & remanet semper A B, hoc est, *x* ∞ *c*. Quod est propositum.

C A P V T II.

POrrò secundo casu, supra expresso, cùm nempe in æquatione, ad simplicissimos terminos reductâ, quantitatum incognitarum altera ad quadratum ascendit, altera verò non item, sed neque in se, neque in alteram quantitatem incognitam ducta reperitur: poterit æquatio ad aliquam sequentium formularum reduci.

$$
\left.\begin{array}{ll}
\text{I.} & yy \infty \; ax \\
\text{II.} & yy \infty \; ax + bb \\
\text{III.} & yy \infty \; ax - bb \\
\text{IV.} & yy \infty -ax + bb
\end{array}\right\} \text{vel conversim} \left\{\begin{array}{l}
ay \infty xx \\
ay + bb \infty xx \\
ay - bb \infty xx \\
bb - ay \infty xx.
\end{array}\right.
$$

Supponendo *y* & *x* esse quantitates incognitas, vel ab initio conceptas, vel postmodum assumptas, ut mox latiùs explicabitur.

T H E O R E M A VII.

Propositio 7.

Si æquatio sit *yy* ∞ *ax*, vel conversim *ay* ∞ *xx* : erit Locus quæsitus Parabola.

Sit ipsius *x* initium immutabile punctum A, atque eadem illa *x* per rectam A B indefinitè se extendere intelligatur, & sit datus vel assumptus angulus æqualis angulo A B C; Assumatur primò eadem A B ut Parabolæ diameter, ad quam ordinatim applicatæ faciant cum ipsa angulos æquales dato vel assumpto angulo A B C, cujusque latus rectum A F

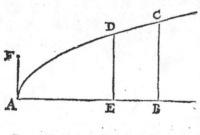

[250]

be equal to the known *a* [2.2]. Then, I say, the required locus is the parabola *ADC'* described through the vertex *A* on the aforementioned diameter and having *a* as the latus rectum corresponding to this diameter.

Indeed, let a point be selected at random on this curve *ADC*, let us say *D*, and let *DE* be drawn at the angle *AED* equal to the given or chosen angle. If *DE* is called *y*, then by the nature of a parabola[2] the square on *ED* is equal to the rectangle *FAE*, which means $y^2 = ax$. Which was proposed.

For a proof of the second part of this theorem, however, with the same hypotheses as before, we have to draw the straight line *AH* from the point *A* parallel to *BC*, and this *AH* has to be chosen as a diameter with which the ordinate-wise applied lines enclose angles equal to the given or chosen angle *ABC* or *AHC*. If the rest of the proof is done as before, then the parabola *ADC* will be the required locus.

Indeed,[3] the square on *GD* or *AE* is equal to the rectangles on *FA* and *AG* or on *FA* and *ED*, respectively. This means $x^2 = ay$. Which was to be demonstrated.

THEOREM VIII

Proposition 8

If the equation is $y^2 = ax + b^2$ or conversely $ay + b^2 = x^2$, then the required locus will be a parabola.

Let point *A* be the immutable initial point of *x* and let us suppose that this *x* extends indefinitely along the straight line *AB* and let the given or chosen angle be equal to the angle *ABC*. Let next *AB* be produced toward *A* until *G*, so that $AG = b^2 / a$. If then *GB* is chosen as a diameter with which the ordinate-wise applied lines enclose angles equal to the given or chosen angle *ABC* and whose latus

---

[1] Lib. I, Prop.1, Cor. 10, p. [168] and Lib. I, Prop. 2, Cor. 4, p. [176].

[2] Lib. I, Prop.1, p. [162]   [3] Lib. I, Prop.1, p. [162]

250   ELEM. CVRVARVM

[margin:]¹ per 10 Coroll. primi, & 4 Coroll. secundi hujus.

sit æquale *a* cognitæ. Dico Parabolam A D C, quæ ¹ per prædictæ diametri verticem A descripta sit, habeatque latus rectum
eidem diametro correspondens ∞ *a*, esse Locum quæsitum.

Sit enim in eadem curva A D C assumptum punctum utcunque, veluti D, ductâque D E in angulo A E D dato vel assumpto
æquali, si ipsa D E vocetur *y*, erit, ex natura Paraboles ² quadratum ex E D ∞ F A E rectangulo, hoc est, *yy* ∞ *ax*. Quod erat
propositum.

[margin:]² per I primi hujus.

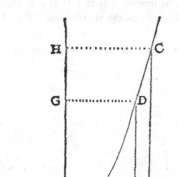

Ad demonstrationem autem secundæ hujus Theorematis partis
iisdem ut supra suppositis, ducenda
est ex A puncto recta A H ipsi B C
parallela, atque eadem A H assumenda pro diametro, ad quam ordinatim applicatæ faciant angulos,
dato vel assumpto angulo A B C
seu A H C æquales, ac cætera, ut
supra, eritque Parabola A D C
Locus quæsitus.

Est enim ³ quadratum ex G D
sive A E quadratum æquale rectangulo sub F A & A G, seu F A
& E D, id est, *xx* ∞ *ay*. Quod erat
demonstrandum.

[margin:]³ per I primi hujus.

THEOREMA VIII.

Propositio 8.

Si æquatio sit *yy* ∞ *ax* + *bb* aut conversim *ay* + *bb*
∞ *xx*, erit Locus quæsitus linea Parabolica.

Sit ipsius *x* initium immutabile A punctum, atque eadem illa *x* per rectam A B indefinitè se extendere intelligatur, sitque
angulus datus vel assumptus æqualis angulo A B C. Deinde producatur A B versùs A usque ad G, ita ut sit A G ∞ $\frac{bb}{a}$; assumptâ-
que G B pro diametro, ad quam ordinatim applicatæ faciant angulos æquales dato vel assumpto angulo A B C, cujusque latus
rectum

[251]

rectum *GF* is equal to the known *a*, then, I say, the required locus is the parabola *GCD*, which is described through the vertex *G* on the aforesaid diameter, which has *a* as latus rectum corresponding to this diameter. Indeed, let a point be selected at random on this diameter, let us say *D*, and let *DE* be drawn at the angle *AED* equal to the given or chosen angle. If *DE* is called *y*, then $y^2 = ax + b^2$, because *GE*, or *AE* + *AG*, equals $x + (b^2 / a)$, and because, by the nature of a parabola,[1] the square on *ED* equals the rectangle on *FG* and *GE*. Which was to be demonstrated in the first place.

To explain, however, the second part of this theorem, with the same hyptheses as before, let the straight line *AH* be drawn from *A* parallel to *BC* and let it be produced toward *A* up to *G* so that $AG = b^2 / a$. If a parabola is described with *GH* as a diameter and with latus rectum *GF* equal to *a*, which parabola intersects the straight line *AB* at *I* then, I say, the curve *ID* is the required locus.

Indeed,[2] by the nature of a parabola, the rectangle enclosed by *FG* and *GH*

---

[1] Lib. I, Prop. 1, p [162]. [2] Lib. I, Prop. 1, p [162].

L I B. II. C A P. II.   251

rectum G F sit æquale *a* cognitæ: dico Parabolam G C D, quæ

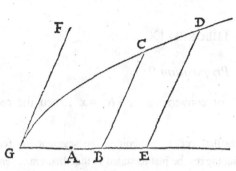

per prædictæ diame-
tri verticem G de-
scripta sit , habeat-
que latus rectum ei-
dem diametro cor-
respondens ∞ *a*, esse
Locum quæsitum.

Sumpto enim in
eadem curva puncto
utcunque , veluti D,
ductâque D E in an-
gulo A E D, dato vel

assumpto æquali, si ipsa D E vocetur *y*, quoniam G E sive A E +
A G est ∞ $x + \frac{bb}{a}$, atque ex natura Paraboles [1] quadratum ex [1] *per* [1] *primi hu-*
E D ∞ rectangulo sub F G & G E, erit *yy* ∞ $ax + bb$. Quod *jus.*
primò erat demonstrandum.

Ad explicationem verò secundæ hujus Theorematis partis iisdem ut supra positis, ducatur ex A recta A H ipsi B C parallela;
eâdemque productâ versùs A usque ad G, ita ut A G sit∞ $\frac{bb}{a}$, di-

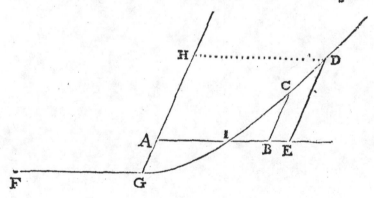

co, si ad G H diametrum latere recto G F ∞ *a* Parabola describatur ut G C, quæ secet rectam A B in I , curvam I D esse Locum
quæsitum.

Est enim [2] ex natura Paraboles rectangulum sub F G & G H *[2] per* [1] *primi hu-*
I i 2   con- *jus.*

[252]

is equal to the square on *HD* or *AE*. Furthermore, as *GH* or *DE* + *AG*, equals $y + (b^2 / a)$ and *FG* = *a*, after due multiplication $ay + b^2 = x^2$. Which was proposed.

THEOREM IX

Proposition 9

If the equation is $y^2 = ax - b^2$ or conversely $ay - b^2 = x^2$, then the required locus will be a parabola.

Let, with the same hypotheses as in the preceding theorem, the segment *AG* (equal to b^2 / a) be taken away from *AB* and let the rest be just as stated in that theorem. Then, I say, *GCD* is the required locus.

Indeed, if we select a point at random on it, let us say *D*, and if we drop *DE* parallel to *CB* and if we call *DE y*, then by the nature of a parabola[1] the square on *ED* or y^2 will be equal to the rectangle on *FG* and *GE*, that is equal to the product of *a* and $x - (b^2 / a)$, namely $ax - b^2$. Which was to be demonstrated and determined.

For an explanation of the second part of this theorem however, with the same hypotheses as before, we draw the straight line *AH* from *A* parallel to *BC* and delete *AG* (equal to b^2 / a). If *GH* is chosen as a diameter and the rest is done as before, then, I say, *GCD* will be the required locus.

---

[1] Lib. I, Prop. 1, p. [162].

252   ÉLEM. CVRVARVM

contentum æquale quadrato ex H D ſeu A E, ac proinde, quo-
niam G H, ſive D E + A G, $\infty\, y + \frac{bb}{a}$, atque F G ∞a, erit, factâ
debitâ multiplicatione, $ay + bb \infty xx$. Quod eſt propoſitum.

THEOREMA IX.

Propoſitio 9.

Si æquatio ſit $yy \infty ax - bb$ aut converſim $ay - bb$
$\infty\, xx$, erit Locus quæſitus linea Parabolica.

Suppoſitis iiſdem, quæ in præcedenti Theoremate, auferatur
ab A B recta A G $\infty \frac{bb}{a}$, fiantque cætera, ut ibidem dictum eſt:
dico curvam G C D eſſe Locum quæſitum.

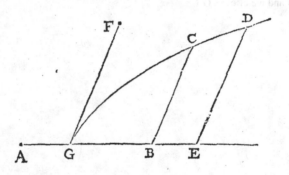

Sumpto enim in ea puncto utcunque, veluti D, demiſsâque
D E ipſi C B parallelâ, ſi eadem D E vocetur y, erit ex natura
Paraboles [1] quadratum ex E D ſeu yy æquale rectangulo ſub
F G & G E, id eſt, producto ex a in $x - \frac{bb}{a}$, nimirum, $ax - bb$.
Quod demonſtrandum determinandumque erat.

Ad explicationem autem ſecundæ hujus Theorematis partis,
iiſdem ut ſupra poſitis, ducatur ex A recta A H ipſi B C parallela,
atque ab ea ſubductâ A G $\infty \frac{bb}{a}$, ſumatur G H pro diametro, &c.
ut ſupra, dico curvam G C D fore Locum quæſitum.

1 per I
primi hu-
jus.

Eſt

[253]

Indeed,[1] by the nature of a hyperbola the rectangle enclosed by *FG* and *GH* is equal to the square on *HD* or *AE*. Furthermore, as *GH* or $DE - AG$ equals $y - b^2 / a$ and $FG = a$, after due multiplication the result is $ay - b^2 = x^2$. Which was proposed.

THEOREM X

Proposition 10

If the equation is $y^2 = b^2 - ax$ or conversely $b^2 - ay = x^2$, then the required locus will be a parabola.

Indeed, let as before point *A* be the immutable initial point of *x* and let us suppose that this *x* extends indefinitely along the straight line *AB* from *A* toward *B*. The given or chosen angle, however, must be equal to the angle *ABC*. Next we measure *AG* (equal to b^2/a) from *A* toward *B* and we choose *GA*

---

[1] Lib. I, Prop. 1, p. [162].

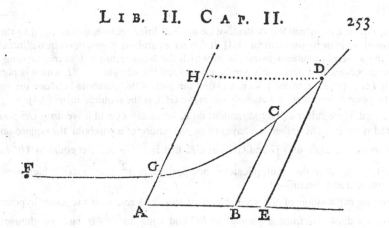

Est enim ¹ ex natura Paraboles rectangulum sub F G & G H ¹ *per ean-* contentum æquale quadrato ex H D seu A E, ideoque, quoniam *dem.*

G H sive D E — A G æquatur $y - \frac{bb}{a}$, atque F G ∞ *a*, erit, factâ debitâ multiplicatione, $ay - bb \infty xx$. Quod erat propositum.

T H E O R E M A X.

Propositio 10.

Si æquatio sit $yy \infty bb - ax$ aut conversim $bb - ay$ ∞xx, erit Locus quæsitus linea Parabolica.

Sit enim, ut supra, ipsius *x* initium immutabile A punctum, intelligaturque eadem *x* in recta A B indefinitè se ab A extendere versùs B ; angulus verò datus vel assumptus esto æqualis angulo

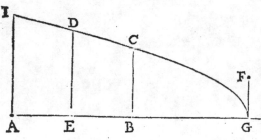

A B C. Deinde ab A versùs B assumptâ A G $\infty \frac{bb}{a}$ sumatur G A pro

[254]

as a diameter with which the ordinate-wise applied lines enclose angles equal to the given or chosen angle or its supplement [2.3]. If further a parabola is described through the vertex G of the aforesaid diameter toward A of which the latus rectum GF corresponding to this diameter is equal to a, and if this parabola intersects the straight line AI, which is parallel to BC, at I (fig. p. [253], transl.), then, I claim: the part of this parabola between the vertex G and the point of intersection I, namely the curve GCI, is the required locus [2.4].

Indeed, if we select a point at random on it, let us say D, and if we drop DE parallel to CB and if we call DE y, then we have: as by the nature of a parabola[1] the square on DE is equal to the rectangle on FG and GE and GE, that is $AG - AE$, is equal to $(b^2 / a) - x$ and $FG = a$, after due multiplication, one arrives at $y^2 = b^2 - ax$. Which was to be demonstrated and determined.

For an explanation of the second part of this theorem, with the same hypotheses as before, we draw AG from A parallel to BC and equal to b^2 / a and we choose GA as diameter. Let the rest be as before with the exception that the point I of intersection lies on the straight line AE.

Let DH be drawn parallel[2] to AB. As by the nature of a parabola the rectangle enclosed by FG and GH is equal to the square on HD or AE, and as GH, that is $AG - ED$, is equal to $(b^2 / a) - y$ and $FG = a$, we have after due multiplication $b^2 - ay = x^2$. Which was proposed.

---

[1] Lib. I, Prop. 1, p. [162].

[2] By the same Proposition.

254   E L E M .   C V R V A R V M

pro diametro , ad quam ordinatim applicatæ faciant angulos
æquales dato vel aſſumpto A B C , aut ejuſdem ad duos rectos
ſupplemento. Quo facto, ſi per prædictæ diametri verticem G
versùs A Parabola deſcribatur , cujus latus rectum G F eidem diametro correſpondens ſit ∞ *a*, quæque Parabola rectam
A I ipſi B C parallelam ſecet in I : dico ejuſdem Parabolæ por-
tionem, inter verticem G & punctum interſectionis I interce-
ptam, nempe curvam G C I, eſſe Locum quæſitum.

Sumpto enim in ea puncto utcunque, veluti D, demiſsâque
D E ipſi C B parallelâ, ſi eadem D E vocetur *y*, cum [1] ex natura Paraboles quadratum ipſius D E ſit æquale rectangulo ſub F G
& G E, & G E ſive A G — A E ſit ∞ $\frac{bb}{a}$ — *x*, ac F G ∞ *a*, factâ
debitâ multiplicatione, erit *y y* ∞ *b b* — *a x*. Quod demonſtrandum determinandumque erat.

*per 1
primi
hujus.*

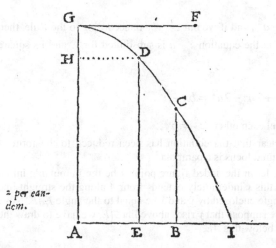

Ad explicationem au-
tem ſecundæ hujus Theo-
rematis partis, iiſdem ut
ſupra poſitis, ex A duca-
tur A G ipſi B C parallela
atque ∞ $\frac{bb}{a}$, aſſumaturque
G A pro diametro , &c.
per omnia, ut ſupra, excepto quòd punctum inter-
ſectionis I ſit in recta A E.

Cum enim ductâ D H
ipſi A B parallelâ [2] ex natura Paraboles rectangu-
lum ſub F G & G H con-
tentum ſit æquale quadra-
to ex H D ſeu A E, ſitque

*per ean-
dem.*

G H ſive A G — E D ∞ $\frac{bb}{a}$ — *y*, atque F G ∞ *a*, factâ multiplicatione, ut decet, erit *b b* — *a y* ∞ *x x*. Quod erat propoſitum.

Regula

[255]

General Rule and method of reducing all equations that result from a suitable operation (when the required locus is a parabola) to one of the four cases that have already been explained in the four preceding theorems

If it happens that the unknown quantity which in the equation is raised to the second power is also found therein in the first power, forming a product with another quantity, either known or unknown, or even with each of them, then we have to replace this quantity with a new one. The new quantity either exceeds the old one or falls short by half the quantity with which the unknown quantity formed a product as mentioned, depending respectively on the plus or minus sign of said product. By means of this the equation itself will be reduced to one of the four preceding cases so that it is not difficult to determine the corresponding parabola with the help of the preceding explanation [2.5].

Examples of the reduction of equations to the form of Theorem VII

If the equation is $y^2 + 2ay = bx - a^2$, and if we put $z = y + a$, according to the Rule, then $z - a = y$. Hence, if everywhere in the equation $z - a$ is substituted for y, and its square for y^2, then we will have:

$$z^2 - 2az + a^2 + 2az - 2a^2 = bx - a^2,$$

that is, omitting all terms that cancel each other, $z^2 = bx$.

From this it is immediately clear that the equation has been reduced to the form of Theorem VII and that thus the required locus is a parabola.

For its accurate determination, let in the added figure point A be the immutable initial point of x, and let us suppose that this x indefinitely extends from A along the straight line AE. And let the given or chosen angle enclosed by y and x be equal to the angle EAF. Now, as $z = y + a$, (transl. we have:) if we suppose that y rises above line AE, we have to draw the straight line GB parallel to AE and below this AE,

L I B. II. C A P. II.   255

Regula univerſalis , moduſque reducendi
omnes æquationes, quæ ex convenienti ope-
ratione producuntur , cùm Locus quæſitus
eſt Parabola , ad aliquem quatuor ca-
ſuum, præcedentibus totidem Theorema-
tibus jam explicatorum.

Si contingat ut quantitas incognita , quæ in æquatione
ad duas dimenſiones aſcendit, in eadem quoque inveniatur unius dimenſionis, cum alia, ſive cognita, ſive incognita quantitate , vel etiam cum utraque planum aliquod faciens, loco ejuſdem aſſumenda eſt alia , vel ipſam excedens, vel ab ea deficiens dimidio quantitatis, quacum illa
planum, uti dictum eſt, conſtituere reperitur, pro diverſa
dicti plani ſigno ✛ vel — affectione. Quo opere ipſa æ-
quatio ad aliquem quatuor præcedentium caſuum redu-
cetur, ita ut ei convenientem lineam Parabolicam determinare, per ea quæ ſuperiùs ſunt explicata, haud difficile ſit.

Exempla reductionis æquationum ad formulam
Theorematis VII.

Si æquatio ſit $yy + 2\,ay \infty\, bx - aa$; aſſumpto, juxta Regulam, $z \infty y + a$, erit $z - a \infty y$. Hinc ſi ubique in æquatione loco
ipſius y ſubſtituatur $z - a$, ejuſdemque quadratum loco yy: habebitur $zz - 2\,az + aa, + 2\,az - 2\,aa \infty bx - aa$, hoc eſt,
omiſſis iis quæ ſeſe mutuò tollunt, erit $zz \infty bx$. Vnde ſtatim
apparet æquationem eſſe reductam ad formulam Theorematis VII, ac proinde Locum quæſitum eſſe Parabolam. Ad cujus
ſpecificam determinationem eſto in appoſita figura ipſius x initium immutabile A punctum, eademque x intelligatur ſe ab A
per rectam A E indefinitè extendere ; ſitque datus vel aſſumptus
angulus, quem y & x comprehendunt, æqualis angulo E A F.
Deinde, quoniam z eſt $\infty y + a$, ſi y ſupra lineam A E exſurgere
intelligatur, ducenda eſt infra eam recta G B ipſi A E parallela,
ita

[256]

in such a way that the part of the straight line *AF* and of all its parallels intercepted between *AE* and *GB* are equal to the known *a*, just as *AG*. Furthermore the aforementioned *GB* has to be chosen as a diameter of the parabola. If on this diameter a parabola is described through its vertex *G*, if the latus rectum *GF* corresponding to this diameter equals *b* and if the parabola intersects *AE* in *I*, then, I say, the curve *ID*, when indefinitely produced toward *D* is the required locus.

Indeed, if we select a point at random on this curve, let us say *D*, and if we draw *DE* parallel to *AF*, and if we call this *DE y*, and if we produce *DE* until it intersects the aforesaid diameter *GB* at *B*, then by construction the intercepted *EB* equals *a*, and so the whole *DB* equals *y* + *a*, that is *z*. Because by the nature of a parabola the square on *DB* equals the rectangle on *FG* and *GB*, or on *FG* and *AE*, we have also $z^2 = bx$ or, if we substitute again $a + y$ for *z*: $y^2 + 2ay + a^2 = bx$, that is $y^2 + 2ay = bx - a^2$. Which was to be demonstrated and determined.

If, however, the equation had been $y^2 - 2ay = bx - a^2$, and if we had applied the same substitution and operation prescribed by the Rule as before, then we would have arrived at the same equation, namely $z^2 = bx$.

But because in this case according to the Rule *z* had to be chosen equal to $y - a$, the diameter *GB* would, therefore (with the same suppositions as before), have fallen   not below the straight line *AE*, but above it and all the rest should have to be performed in the same way as before.

If, however, the equation is $by - a^2 = x^2 + 2ax$, which is the converse of the equation explained before (p. [255], transl.), and if we assume $v = x + a$, according to the Rule, then $v - a = x$. Hence, if in the equation we substitute $v - a$ for *x* and

ita ut pars rectæ A F, omniumque ipsi parallelarum, intercepta inter A E & G B, veluti A G, æquetur *a* cognitæ. Porrò prædicta G B assumenda est ut Parabolæ diameter, ad quam si per ejusdem verticem G, existente G F latere recto, ipsi diametro G B correspondente, ∞ *b* Parabola describatur, secans rectam A E in I : dico curvam I D indefinitè versùs D productam esse Locum quæsitum.

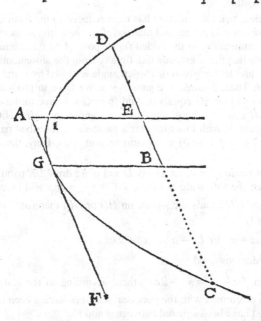

Etenim assumpto in eadem curva puncto utcunque, veluti D, ductâque D E ipsi A F parallelâ, si eadem D E vocetur *y*, producaturque donec prædictæ diametro G B occurrat in *B* : erit ex constructione intercepta E B ∞ *a*, ac proinde tota D B ∞ *y + a*, hoc est, *z*. Quare cum ex natura Paraboles quadratum ex D B æquetur rectangulo sub F G & G B, vel F G & A E : erit quoque *z z* ∞ *b x*, sive, restituto *y + a* loco *z*, *yy + 2 ay + aa* ∞ *b x*, id est, *yy + 2 ay* ∞ *b x — a a*. Quod demonstrandum determinandumque erat.

Quòd si æquatio fuisset *yy — 2 ay* ∞ *b x — a a*, factâ assumptione secundùm Regulam, atque operatione, ut supra; deventum fuisset ad eandem æquationem, nimirum, *z z* ∞ *b x*. Sed quoniam z eo casu juxta Regulam assumenda fuisset ∞ *y — a*, idcirco quoque diameter G B (iisdem ut supra positis) non infra, sed supra rectam A E cecidisset, cæteraque omnia eodem quo supra modo expedienda fuissent.

Si verò æquatio sit *b y — a a* ∞ *x x + 2 a x*, quæ est conversa superiùs exposita, assumpto juxta Regulam *v* ∞ *x + a*, erit *v — a* ∞ *x*. Quare si loco ipsius *x* in æquatione substituatur *v — a*, atque hujus

[257]

its square for x^2, then $by - a^2 = v^2 - 2av + a^2 + 2av - 2a^2$, which means $by = v^2$, if we omit the terms that cancel each other.

From this it is immediately clear that the equation has been reduced to the form of the converse of the aforementioned seventh Theorem and that therefore the required locus is a parabola. Let, for its accurate determination, in the added figure point A be the immutable initial point of x and let us suppose that this x extends indefinitely from the aforementioned point A along the straight line AE and let the given or chosen angle enclosed by y and x, be equal to the angle AGH or FGH. Then, because v equals $x + a$, we have to produce the straight line AE toward A up to G so that AG equals a. And from G we have to draw GH enclosing the angle EGH or FGH equal to the given or chosen angle and GH has to be chosen as a diameter of the parabola. If with this diameter a parabola is described passing through G as its vertex and with FG (equal to b) as its latus rectum, then, I say, the curve GD is the required locus.

Indeed, if we select a point at random on it, let us say D and if we draw DE parallel to HG and if we call this DE y, then the following holds: as $GE = x + a$ or v and as by the nature of a parabola the rectangle FGH equals the square on HD or GH, therefore $by = v^2$ or, if we substitute again $x + a$ for v,

$$by = x^2 + 2ax + a^2 \text{ or } by - a^2 = x^2 + 2ax.$$

Which was to be determined and demonstrated.

But if the equation had been $by - a^2 = x^2 - 2ax$, then, according to the Rule, the same operation should have been performed with the necessary changes having been made. And in this case the point G would have been situated between A and E.

Lib. II. Cap. II.   257

hujus quadratum loco *x x*: erit *b y* — *a a* ∞ *v v* — 2 *a v* + *a a*, + 2 *a v* — 2 *a a*, hoc est, omissis iis, quæ se mutuò tollunt, erit *b y* ∞ *v v*.

Vnde statim apparet, reductam esse æquationem ad formulam prædicti Theorematis septimi conversim, ac proinde Locum quæsitum esse Parabolam. Ad cujus specificam determinationem esto in apposita figura ipsius *x* initium immutabile punctum A,

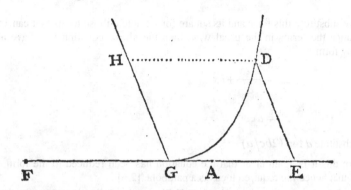

intelligaturque eadem *x* à prædicto puncto A per rectam A E indefinitè se extendere, sitque datus vel assumptus angulus, quem comprehendunt *y* & *x*, æqualis angulo A G H vel F G H. Deinde, quoniam *v* æquatur *x* + *a*, producenda est recta A E versùs A usque ad G, ita ut A G sit ∞ *a*; & ex G ducenda est G H, faciens angulum E G H vel F G H dato vel assumpto angulo æqualem, ipsaque G H sumenda est pro Parabolæ diametro, ad quam si per ejus verticem G atque latere recto F G ∞ *b* Parabola describatur, ut G D: dico curvam G D esse Locum quæsitum.

Sumpto enim in ea puncto utcunque, veluti D, ductâque D E ipsi H G parallelâ, si eadem D E vocetur *y*, cum G E sit ∞ *x* + *a* seu *v*, atque ex natura Paraboles F G H rectangulum ∞ quadrato ex H D sive G E, erit *b y* ∞ *v v*, sive, restituto *x* + *a* loco *v*, *b y* ∞ *x x* + 2 *a x* + *a a*, seu *b y* — *a a* ∞ *x x* + 2 *a x*. Quod determinandum, demonstrandumque erat.

Quòd si æquatio fuisset *b y* — *a a* ∞ *x x* — 2 *a x*, eadem per omnia mutatis mutandis secundùm Regulam instituenda fuisset operatio, cecidissetque eo casu punctum G inter A & E.

Pars II.   K k   Eodem

[258]

In the same way the following holds. If the equation is

$$y^2 + \frac{2bxy}{a} + 2cy = bx - \frac{b^2x^2}{a^2} - c^2$$

and if we put according to the Rule

$$z = y - \frac{bx}{a} + c,$$

then

$$y = z - \frac{bx}{a} - c.$$

If we substitute this for y and its square for y^2, delete those terms that cancel each other and arrange the terms in the usual way, then the above equation will have assumed the following form

$$z^2 = \frac{2bcx}{a} + bx$$

or

$$z^2 = dx,$$

if we substitute d for $(2bc/a) + b$.

From this it is again clear that the equation has been reduced to the form of Theorem VII and that hence the required locus is a parabola. [2.6] .

Let, for its accurate determination [2.7], in the following figure point A be the immutable initial point of x and let us suppose that this x extends indefinitely from the point A along the straight line AE. And let the given or chosen angle, enclosed by y and x, be equal to the angle EAF or EAG. Then, as $z = y + c + (bx/a)$, (supposing that y rises above the line AE, just as ED,) we first have to draw the straight line GB below it (that is AE, transl.) and parallel to it, so that the parts of the straight line FG and of all its parallels intercepted between the aforesaid AE and GB are equal to the known c, just as AG and EB. Hereafter the following holds: because every segment that can act as y, will become $y + c$, when produced up to the straight line GB just as for instance DB, we still must add bx/a to it in order to obtain the assumed z.

Therefore we proceed as follows: because GB or AE equals x, when assumed to be indefinite, according to Theorem I of this book (p. [244], transl.) we have to draw a straight line below this GB from G, e.g. GC, so that the segments of all parallels to GF intercepted between GB and GC, e.g. BC, have the same ratio to the segments of GB intercepted between G and the aforementioned parallels,

258   E L E M.   C V R V A R V M

Eodem modo si æquatio sit $yy + \frac{2bxy}{a} + 2cy \infty bx - \frac{bbxx}{aa} - ec$,

assumpto juxta Regulam $z \infty y + \frac{bx}{a} + c$: erit $y \infty z - \frac{bx}{a} - c$.

Quo substituto in locum ipsius y, ejusdemque quadrato loco yy, expunctisque iis, quæ se invicem tollunt, atque omnibus rite ordinatis sequentem formam induta erit superior æquatio:

$zz \infty \frac{2bc}{a} x + bx$, aut $zz \infty dx$, si loco $\frac{2bc}{a} + b$ substituatur d.

Vnde iterum apparet, æquationem esse reductam ad formulam Theorematis VII, ac propterea Locum quæsitum esse Parabolam. Ad cujus specificam determinationem esto in sequenti figura ipsius x initium immutabile punctum A, atque eadem x ab A puncto per rectam A E indefinitè se extendere intelligatur, sitque datus vel assumptus angulus, quem y & x comprehendunt, æqualis angulo E A F vel E A G. Deinde quoniam z est $\infty y + c + \frac{bx}{a}$,

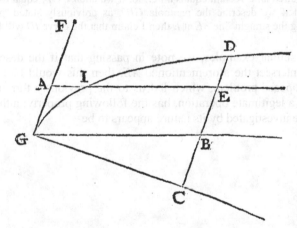

si y supra lineam A E exsurgere intelligatur, veluti E D, ducenda primùm est infra eandem recta G B ipsi parallela, ita ut partes rectæ F G omniumque ipsi æquidistantium inter prædictas A E & G B interceptæ, veluti A G, E B, æquentur c cognitæ. Quo peracto, cum quævis recta, quæ possit esse y, ad rectam G B producta, ut, exempli gratiâ, D B, sit $\infty y + c$, oportet ipsi adhuc adjungere $\frac{bx}{a}$, ut fiat æqualis z assumptæ.

Quare, cum G B seu A E indefinitè sumpta sit ∞x, si ex G juxta I Theorema hujus libri infra eandem G B recta ducatur, ut G C; ita ut omnium ipsi G F parallelarum partes inter G B & G C interceptæ, veluti B C, ad partes ipsius G B inter G & dictas parallelas
rallelas

[259]

like *BG* namely the ratio of *b* to *a* [2.8].

To this end, let us arrange that *GB* is to *BC* as *a* is to *b*, whence $BC = bx \,/\, a$. By the same argument all segments parallel to *BC* that are drawn from *GB* up to *GC*, will equal $bx \,/\, a$. And so any segment rising above *AE* that can act as *y* will be equal to $y + c + bx \,/\, a$ or *z* after it has been produced up to *GC* as for example *DC*.

As its square has to be equal to *dx*, the following, then, is immediately clear. If we describe a parabola with *GC* as a diameter, whose latus rectum *GF* has been chosen so that the rectangles on this latus rectum and the segments of the diameter intercepted between the vertex and the ordinate-wise applied lines are equal to *dx*, then this parabola will be the required locus.

However, the ratio of the segment *GB* to the segment *BC* and the ratio of other similar segments is known, namely as *a* to *b*, and the angle *GBC* enclosed by them is   known as well; it is namely equal to the given or chosen *EAF*. Therefore the ratio of *GB* to *GC* and the ratio of other similar segments is known as well.[1] Let it be as the known *a* to a known *e*. Therefore, as *GB* or *AE*, assumed to be indefinite, may be expressed by *x*, *GC*, likewise assumed to be indefinite, that is every part of the diameter intercepted between the vertex and the ordinate-wise applied lines, equals *ex/a*. After multiplication by the latus rectum this must yield the term *dx* of the equation. Therefore the same term *dx* of the equation, when divided by *ex/a*, must yield the aforementioned latus rectum as. And so we learn by this division that the required latus rectum equals *ad/e*. So, if we choose *GF*, equal to *ad/e*, as the latus rectum and if we describe the parabola *GID*, as previously stated, on the diameter *GC*, intersecting the straight line *AE* at *I*, then I claim that the curve *ID* will be the required locus.

Here, as in other similar examples, we note in passing that if the described parabola should not intersect the aforementioned *AE*, then this would be a sure indication that the proposed problem, which led us to the previously formulated equation by means of a legitimate operation, has the following property: although the locus that has to be investigated by its nature appears to be

---

[1] VI, 6.

rallelas interceptas, veluti B G, eandem rationem habeant, quæ
est inter b & a. Quod ipfum ut fiat, ftatuatur ut a ad b, ita G B ad
B C: eritque B C ∞ $\frac{bx}{a}$. Eodem modo rectæ omnes ipfi B C parallelæ, quæ à G B ad G C ducuntur, erunt ∞ $\frac{bx}{a}$. Atque ita recta quælibet fupra A E exfurgens, quæ poffit effe y, poftquam ad
rectam G C erit producta, ut, exempli gratiâ, D C, erit ∞ $y + c$
$+ \frac{bx}{a}$ feu z. Hujus igitur quadratum cum debeat effe ∞ dx, ftatim inde apparet, fi Parabola defcripta foret ad diametrum G C,
cujus latus rectum G F ita effet affumptum, ut rectangula, fub
codem latere recto & diametri portionibus, inter verticem & ordinatim applicatas interceptis, contenta, forent ∞ dx, eandem
illam Parabolam fore Locum quæfitum. At verò cum ratio rectæ G B ad rectam B C, aliarumque fimilium, cognita fit, nempe, ut a ad b; fitque itidem notus angulus G B C, fub iifdem comprehenfus, utpote æqualis dato vel affumpto E A F: erit propterea quoque ¹ nota ratio G B ad G C, aliarumque fimilium, ₁ *per 6.*
quæ fit ut a cognitæ ad e cognitam. Hinc cum G B feu A E inde- *fexti.*
finitè fumpta exprimatur per x, erit G C itidem indefinitè fumpta,
hoc eft, omnis diametri portio inter verticem & ordinatim applicatas intercepta ∞ $\frac{ex}{a}$. Quæ cum in latus rectum ducta producere debeat æquationis terminum dx, idem quoque æquatio-
nis terminus dx per $\frac{ex}{a}$ divifus ut prædictum latus rectum refti-
tuat neceffe eft: ac proinde per eandem divifionem cognofcitur
quæfitum latus rectum æquari $\frac{ad}{e}$. Sumptâ ergo G F ∞ $\frac{ad}{e}$ pro
latere recto, fi ad diametrum G C, ut fupra dictum eft, defcri-
batur Parabola G I D, fecans rectam A E in I: dico curvam I D
fore Locum quæfitum.

Atque hîc, ut & in aliis fimilibus exemplis obiter
notandum, fi Parabola defcripta prædictam A E non
fecaret, id certo indicio fore, quæftionem propofitam,
per quam legitimâ operatione ad fupra expreffam æ-
quationem perventum fuerit, ejus effe conditionis, ut
Locus ad indagandum propofitus fui quidem naturâ
linea

[260]

a parabola, nevertheless no parabola can be described as a solution to the problem because the given quantities cannot be matched in the required way [2.9].

To demonstrate, the foregoing [2.10], however, let us select a point at random on the curve ID, say D, and let us draw DE parallel to FG, which (that is ID, transl.), being produced, intersects the straight line GB at B and meets the diameter GC at C (fig. on p. [258], transl.). If we call $DE\,y$, then the whole DC equals $y + c + (bx/a)$, that is z, because EB or AG equals c and as $BC = (bx/a)$. Since by the nature of a parabola the square on DC equals the rectangle FGC, we also have, on the basis of the foregoing, $z^2 = dx$. So if we substitute or restore $y + c + (bx/a)$ for z and likewise $(2bc/a) + b$ for d and if we delete the terms that cancel each other because they are equal, and arrange all terms as it should be, then

$$y^2 + \frac{2bxy}{a} + 2cxy = bx - \frac{b^2x^2}{a^2} - c^2 .$$

Which was to be determined and demonstrated.

If, however, the equation had been $y^2 - (2bxy/a) - 2cxy = bx - (b^2x^2/a^2) - c^2$, then we would have arrived at the same equation (that is $z^2 = dx$, transl.) with the assumptions of the Rule and by a suitable operation. But because in this case according to the assumptions we had to put $z = y - (bx/a) - c$, according to the same hypotheses as before, the straight line GB would not have to be drawn below the straight line AE but above it, just as also GC would not have to be drawn below this GB, but above it. All the rest would have to be performed in the same way.

If the equation is $by - (b^2y^2/a^2) - c^2 = x^2 + (2byx/a) + 2cx$, which is the counterpart of the previously explained equation, and if we put according to the Rule $v = x + (by/a) + c$, then $x = v - (by/a) - c$. Hence, if we substitute this value for x and its square for x^2, and if we delete the terms that cancel each other, and arrange all terms as usual, then the previous equation will have the following form: $(2bc/a)y + by = v^2$, or (if we substitute d for $(2bc/a) + b$), $dy = v^2$. This once more shows that the proposed equation has been reduced to the converse form of the aforementioned Theorem VII and, therefore, that the required locus is a parabola.

260   E L E M.   C V R V A R V M

linea Parabolica exiſtat; ſed quòd nulla tamen quæſtioni ſatisfaciens deſcribi poſſit, cum propoſitæ quantitates, eo, ut petitur, modo, conjungi nequeant.

Ad demonſtrationem autem eorum, quæ ſupra diɕta ſunt, ſumatur in curva I D punɕtum utcunque, veluti D, duɕtâque D E ipſi F G parallelâ, quæ protraɕta ſecet reɕtam G B in B, occurratque diametro G C in C, ſi D E vocetur y, cum E B ſeu A G ſit $\infty\ c$, & B C $\infty\ \dfrac{bx}{a}$, erit tota D C $\infty\ y + c + \dfrac{bx}{a}$, hoc eſt, z. Cumque ex natura Parabolæ quadratum ex D C ∞ F G C reɕtangulo, erit quoque ex antediɕtis $z\,z \infty\ d\,x$. Ac proinde ſubſtitutis aut reſtitutis $y + c + \dfrac{bx}{a}$ loco z, itemque $\dfrac{2bc}{a} + b$ in locum ipſius d, & ablatis quæ propter æqualitatem ſe invicem tollunt, ordinatiſque omnibus, ut decet, erit $yy + \dfrac{2bxy}{a} + 2\,cy \infty bx - \dfrac{bbxx}{aa} - cc$. Quod determinandum, demonſtrandumque erat.

Sin autem æquatio fuiſſet $yy - \dfrac{2bxy}{a} - 2cy \infty bx - \dfrac{bbxx}{aa} - cc$, faɕtâ aſſumptione ſecundùm Regulam atque operatione uti decet, ad eandem æquationem perventum fuiſſet; ſed quoniam z juxta aſſumptionem eo caſu faciendam fuiſſet æqualis $y - \dfrac{bx}{a} - c$, idcirco quoque ſuppoſitis, ut ante, reɕtâ G B non infra ſed ſupra reɕtam A E, ut & G C non infra ſed ſupra eandem G B ducenda fuiſſet, cæteraque omnia eodem quo ſupra modo fuiſſent expedienda.

Si verò æquatio ſit $by - \dfrac{bbyy}{aa} - cc \infty xx + \dfrac{2byx}{a} + 2cx$, quæ eſt converſa ſuperiùs expoſitæ, aſſumpto juxta Regulam $v \infty x + \dfrac{by}{a} + c$, erit $x \infty v - \dfrac{by}{a} - c$. Vnde ſubſtituto hoc valore in locum ipſius x, ejuſdemque quadrato loco xx, expunɕtiſque iis, quæ ſe invicem tollunt, atque omnibus rite ordinatis, ſuperior æquatio ſequenti formâ induta erit $\dfrac{2bc}{a}y + by \infty vv$, aut (ſi loco $\dfrac{2bc}{a} + b$ ſubſtituatur d) $dy \infty vv$. Id quod rurſus arguit æquationem propoſitam reduɕtam eſſe ad formulam prædiɕti Theorematis V I I converſim, ac proinde Locum quæſitum eſſe Parabolam.

Ad

[261]

Let, for its accurate determination, in the following figure point A be the immutable initial point of x and let us suppose that this x extends indefinitely from point A along the straight line AE and let the given or chosen angle be EAH or FAH. By the second part of Theorem VII it is certain that the aforesaid parabola has to be described in such a way that the segments that are ordinate-wise applied to its diameter are parallel to AE and that, according to the proposed equation, they must be equal to the assumed quantity v, that is to $x + (by/a) + c$. Therefore we must first draw the straight line GB parallel to AH, so that the part of EA, when produced in the direction of A, as well as all its parallel segments, say AG or HB are equal to the known c. Hereafter we proceed as follows. Any segment that can exist parallel and equal to AE and that therefore can be expressed by x, as for instance DH, equals $x + c$ when produced up to the straight line GB, just as DB. Therefore the diameter of the parabola has to be drawn from point G and has to be situated (according to what has been explained before) on the opposite side of GB from point E on the straight line GE (the text erroneously gives GB, transl.) in such a way that if GC, indefinitely assumed, is called y, then BC and the parts of all parallels to AE, which are intercepted between GC and the straight line GB can be expressed by by/a. And so every segment that is parallel to AE and can act as x when produced toward the straight line GC, just as DC, is equal to $x + by/a + c$, that is v. As its square must be equal to the other term of the equation, namely dy [2.11], the following is immediately clear: if a parabola had been described on the diameter GC

L I B. II. C A P. II.   261

Ad cujus fpecificam determinationem efto in fequenti figura
ipfius *x* initium immutabile punctum A, atque eadem *x* à puncto
A per rectam A E indefinitè fe extendere intelligatur, fitque da-

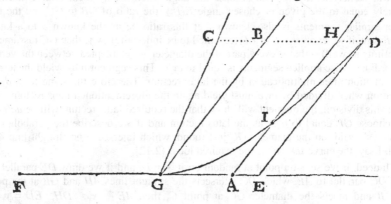

tus vel affumptus angulus E A H vel F A H. Deinde, quoniam
ex fecunda parte Theorematis V I I conftat, prædictam Parabolam ita effe defcribendam, ut ordinatim applicatæ ad ejus diametrum fint ipfi A E parallelæ, debeantque juxta æquationem propofitam æquales effe quantitati affumptæ *v*, hoc eft, $x + \frac{by}{a} + c$,
ducenda primùm eft recta G B ipfi A H parallela, ita ut pars rectæ E A, versùs A productæ, ut & omnium ipfi æquidiftantium,
velut A G vel H B fit ∞ *c* cognitæ. Quo facto, cum quævis recta,
quæ poffit effe ipfi A E æquidiftans & æqualis, ac proinde exprimi-
mi per *x*, ut, verbi gratiâ, D H, ad rectam G B producta; uti D B,
æquetur *x + c*: ita porrò è puncto G ducenda, &, fecundùm ea,
quæ in præcedentibus explicata funt, conftituenda eft Parabolæ
diameter ab adverfa parte ipfius G B, quàm eft punctum E in
recta G C, ut, fi G B indefinitè vocetur *y*, B C, aliarumque
omnium ipfi A E parallelarum inter eandem G C & rectam G B
interceptæ partes exprimantur per $\frac{by}{a}$. Atque ita quælibet recta
ipfi A E parallela, quæ poffit effe *x* ad rectam G C producta, veluti D C, fit ∞ $x + c + \frac{by}{a}$, hoc eft, *v*. Cujus quidem quadratum
cum æquale effe debeat alteri æquationis termino, nempe, *dy*:
ftatim apparet, fi Parabola defcripta foret ad diametrum G C,

Kk. 3   cujus

[262]

whose latus rectum *GF* had been chosen so that the rectangles enclosed by this latus rectum and the parts of the diameter intercepted between the vertex *G* and the ordinate-wise applied lines were equal to *dy*, then this parabola would be the required locus. But because the ratio of the segment *GB* to the segment *BC* and the ratio of other similar segments is known, namely as *a* to *b*, and because the angle enclosed by them is also known (it is namely equal to the given or chosen angle *EHA*), the ratio of *GB* to *GC* and the ratio of other similar segments is also[1] known. Let this ratio be as the known *a* to a known *e*. Therefore, if we express *GB* or *ED* (assumed to be indefinite) by *y*, then *GC* (assumed to be indefinite as well) − that is every part of the diameter − intercepted between the vertex and the ordinate-wise applied segments is equal to *ey/a*. This segment must yield the term *dy* of the equation after multiplication by the latus rectum. Therefore the same term *dy* of the equation when divided by *ey/a* must yield again the aforementioned latus rectum. And so, after this division, the quotient will show that the required latus rectum will be *ad/e*. So, if we choose *GF* equal to *ad/e* as the latus rectum and if we describe the parabola *GID*, as previously said, on the diameter *GC* just found, which intersects the straight line *AH* at *I*, then, I say: the curve *ID* will be the required locus [2.12].

Indeed, if we select a point at random on it, say *D*, and if we draw *DE* parallel to *AH*, and *DC* parallel to *AE*, which *DC* intersects the straight lines *AH* and *GB* at the points *H* and *B* and meets the diameter *GC* at point *C*, then $AE = x = DH$, $ED = y = GB$, $AG = HB = c$, and *BC* =*by/a* and so the whole *DC* satisfies $DC = x + c + (by/a)$, that is *v*. As by the nature of a parabola the rectangle *FGC* is equal to the square on *DC*, therefore $dy = v^2$ after multiplication of *ad/e* by *ey/a* and after multiplication of *v* by itself [2.13]. If we substitute or restore $x + c + by/a$ for *v* and also $2bc/a + b$ for *d*, and delete the terms that cancel each other because they are equal and if we arrange all terms as is suitable, then

$$by - \frac{b^2 y^2}{a^2} - c^2 = x^2 + \frac{2byx}{a} + 2cx.$$

Which was to be determined and demonstrated.

It would, however, be superfluous to go at greater length into the other cases related to the aforementioned form, as these can easily be explained,

---

[1] VI, 6

262   ELEM. CVRVARVM

cujus latus rectum G F ita esset assumptum, ut rectangula contenta sub eodem latere recto & diametri portionibus, inter verticem
G & ordinatim applicatas interceptis, forent ∞ dy, eandem illam
Parabolam fore Locum quæsitum. At verò cum ratio rectæ G B
ad rectam B C aliarumque similium cognita sit, nimirum, ut a
ad b; sitque itidem notus angulus sub iisdem comprehensus, utpote æqualis dato vel assumpto E A H: erit quoque ^{1} ratio ipsius
G B ad G C aliarumque similium cognita, quæ sit ut a cognitæ
ad e cognitam. Quocirca si G B sive E D indefinitè sumpta exprimatur per y, erit G C itidem indefinitè sumpta, hoc est, omnis
diametri portio, inter verticem & ordinatim applicatas intercepta ∞ $\frac{cy}{a}$. Quæ cum in latus rectum ducta producere debeat æ-
quationis terminum dy, idem quoque æquationis terminus dy per
$\frac{ey}{a}$ divisus ut prædictum latus rectum restituat necesse est. ac proinde factâ eâdem divisione indicabit quotiens latus rectum quæ-
situm fore $\frac{ad}{e}$. Hinc, sumptâ G F ∞ $\frac{ad}{e}$ pro latere recto, si ad diametrum G C inventam, ut supra dictum est, describatur Para-
bola G I D, secans rectam A H in I: dico curvam I D fore Locum quæsitum.

Sumpto enim in eadem puncto utcunque, veluti D, ductâque
D E ipsi A H, ut & D C ipsi A E parallelâ, quæ quidem D C secet rectas A H & G B in punctis H & B, occurratque diametro
G C in puncto C: erit A E ∞ x ∞ D H; E D ∞ y ∞ G B; A G &
H B ∞ c; B C ∞ $\frac{by}{a}$ ideoque tota D C ∞ $x+c+\frac{by}{a}$, hoc est, v.
Cumque ex natura Parabolæ rectangulum F G C sit æquale qua-
drato D C: erit, factâ multiplicatione $\frac{ad}{e}$ in $\frac{ey}{a}$, atque v in se
ipsam, dy ∞ vv. Et substitutis aut restitutis $x+c+\frac{by}{a}$ loco v,
itemque $\frac{2bc}{a}+b$ in locum ipsius d, atque ablatis quæ propter æ-
qualitatem se invicem tollunt, ordinatisque omnibus, ut decet,
$by-\frac{bbyy}{aa}-cc \infty xx+\frac{2byx}{a}+2\,cx$. Quod determinandum, demonstrandumque erat.

De cæteris autem casibus, ad prædictam formulam spectanti-
bus, supervacuum fuerit plura exponere, cum ex prædictis facilè
expli-

^{1} per 6
sexti.

[263]

determined, and demonstrated by what we stated before. One has only to observe the various positions of the curves that necessarily result from the different distributions of the plus and minus signs. A second reason is that I will later on explain all cases of similar loci on the basis of a general Rule [2.14].

Examples of the reduction of equations to the form of Theorem VIII

If the equation is

$$y^2 - \frac{bxy}{a} = -\frac{b^2x^2}{4a^2} + bx + d^2 \, ,$$

and if, according to the general Rule, we put

$$z = y - \frac{bx}{2a} \, , \text{ then } y = z + \frac{bx}{2a} \, .$$

If we substitute this for y, and its square for y^2, omit all terms that cancel each other and arrange all terms as usual, then the above equation will have the following form

$$z^2 = bx + d^2 \, .$$

From this it is clear that the equation has been reduced to the form of Theorem VIII and hence the required locus is a parabola.

Let, for its accurate description, in the added figure point A be the immutable ini-tial point of x and let us suppose that this x extends indefinitely from that point A

explicari, determinari, ac demonstrari queant; observatâ solum-
modo diversâ linearum positione, quæ ex signorum + & — dif-
ferentia oriri debet, cumque omnes similium locorum casus mox
per generalem Regulam sim exhibiturus.

Exempla reductionis æquationum ad formulam Theo-
rematis VIII.

Si æquatio sit $yy - \frac{bxy}{a} \infty - \frac{bbxx}{4aa} + bx + dd$, assumpto juxta

Regulam $z \infty y - \frac{bx}{2a}$, erit $y \infty z + \frac{bx}{2a}$. quo substituto in locum

ipsius y, & ejusdem quadrato loco yy, omissisque iis, quæ se invicem tollunt, atque omnibus rite ordinatis, æquatio superior sequenti formâ erit induta: $zz \infty bx + dd$.

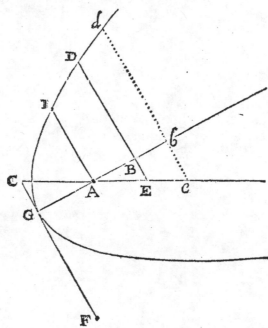

Vnde apparet, eandem esse reductam ad formulam Theorematis VIII, ac proinde Locum quæsitum esse Parabolam. Ad
cujus particularem descriptionem esto in adjunctâ figura ipsius x
initium immutabile punctum A, atque eadem x à dicto puncto A

per

[264]

along the straight line AE from the said point A and let the given or chosen angle, which is enclosed by y and x, be equal to the angle AED. Now $z = y - (bx/2a)$,, so assuming that y rises above line AE (like ED), we have also to draw the straight line AB from point A and above line AE so that the ratio of AE to EB is the same as the ratio of the known $2a$ to the known b. This means that AE or x is to EB as $2a$ is to b and so $EB = bx/2a$. The same considerations hold for all other segments parallel to EB and intercepted between AE and AB. Indeed each of these (segments) will be equal to $bx/2a$. Therefore, as is clear from what we stated before, if the term d^2 were missing in the equation, the said AB would be the diameter of the parabola and point A its vertex. If we put the ratio of AE to AB as $2a$ to e, then the corresponding latus rectum would be $2ab/e$ [2.15]. However, as the rectangle enclosed by the latus rectum and the part of the diameter intercepted between the vertex and the ordinate-wise applied lines has to be equal to $bx + d^2$, the following is clear: if, with the same suppositions, we produce the diameter

per rectam A E indefinitè se extendere intelligatur, sitque datus
vel assumptus angulus, quem y & x comprehendunt, æqualis angulo A E D. Deinde, quoniam $z \infty y - \frac{bx}{2a}$, si y supra lineam A E
exsurgere intelligatur, veluti E D, ducenda quoque est supra lineam A E ex puncto A recta A B, ita ut eadem sit ratio A E ad
E B, quæ est ipsius $2a$ cognitæ ad b cognitam, hoc est, ut sit uti
$2a$ ad b, ita A E seu x ad E B, eritque E B $\infty \frac{bx}{2a}$. idem intellige
de omnibus aliis rectis ipsi E B parallelis, atque inter A E & A B
interceptis, quæ quidem singulæ ipsi $\frac{bx}{2a}$ erunt æquales. Hinc,

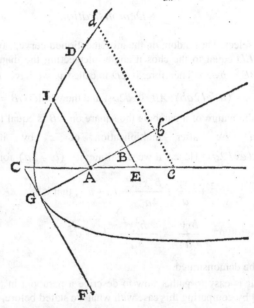

quemadmodum ex supra dictis patet, si terminus dd in æquatione deficeret, prædicta A B Parabolæ diameter foret, ejusque
vertex punctum A, &, positâ ratione A E ad A B, ut $2a$ ad e, latus rectum ipsi correspondens esset $\infty \frac{2ab}{e}$. Iam verò cum rectangulum, quod sub latere recto & portione diametri, inter verticem atque ordinatim applicatas interceptâ, continetur, æquale
esse debeat $bx + dd$: manifestum est, si, iisdem positis, diameter
A B

[265]

AB in the direction of A to G so that the rectangle enclosed by the said latus rectum and the segment GA, equals d^2, then the straight line GB will be the required diameter and its vertex will be the said point G. And so d^2 divided by the said latus rectum (that is by $2(ab/e)$) will be equal to the lenghth of GA. And so GA will be equal to $d^2e/(2ab)$. Therefore, if we describe a parabola GDd with diameter GB, with latus rectum GF equal to $2(ab/e)$ and at the given angle (refers to the latus rectum, transl.), which parabola intersects AI (parallel to ED) at I, then, I say, the curve IDd will be the required locus.

However, let it also be noted here in passing that the said vertex G can also be found in the following way: namely, if we produce EA up to C so that $AC = d^2/b$ and if we next draw CG through point C parallel to DE, intersecting AB (being produced) at G, then, indeed, the required vertex is situated at this point of intersection [2.16].

Demonstration

Let a point be selected at random on the aforementioned curve, say D and let DE be drawn at the angle AED equal to the chosen angle intersecting the diameter GB at B. Then, by construction, $BE = bx/2a$. Therefore, if ED is called y, we have $DB = y - (bx/2a) = z$; $FG = 2ab/e$; $GA = (d^2e/2ab)$; $AB = ex/2a$, and the whole $GB = (d^2e/2ab) + (ex/2a)$.

But as by the nature of a parabola the square on DB is equal to the rectangle FGB, we have $z^2 = d^2 + bx$ after multiplication of z by itself and $2ab/e$ by $(d^2e/2ab) + (ex/2a)$. Hence, if we substitute $y - (bx/2a)$ for z, we will obtain

$$y^2 + \frac{bxy}{a} + \frac{b^2x^2}{4a^2} = bx + d^2 \text{ ; that is}$$

$$y^2 - \frac{bxy}{a} = -\frac{b^2x^2}{4a^2} + bx + d^2 .$$

Which was to be demonstrated.

However, it is easy to gather how to describe a parabola in the converse case of this example [2.17] by comparing this case with what we stated before.

If the equation were

$$\frac{bcy}{a} + by - \frac{b^2y^2}{a^2} + \tfrac{1}{4}c^2 = x^2 + \frac{2byx}{a} - cx ,$$

Lib. II. Cap. II.   265

A B versùs A producatur ad G, ita ut rectangulum sub prædicto latere recto & parte G A contentum sit ∞ dd, rectam G B quæsitam fore diametrum, ejusque verticem prædictum G punctum: ac proinde & dd per prædictum latus rectum, hoc est per $\frac{2ab}{c}$, divisum æquari longitudini G A, ideoque G A fore ∞ $\frac{dde}{2ab}$. Quare si diametro G B & latere recto G F ∞ $\frac{2ab}{c}$ in dato angulo Parabola describatur G D d, secans A I ipsi E D parallelam in I : dico curvam I D d fore Locum quæsitum.

Verùm obiter hîc quoque notandum venit, prædictum verticem G etiam inveniri hoc pacto: si nempe E A producatur ad C, ita ut A C sit ∞ $\frac{dd}{b}$, ac deinde per punctum C ipsi D E parallela ducatur C G, occurrens productæ A B in G: erit enim in eodem illo concursus puncto vertex quæsitus.

Demonstratio.

Sumatur in prædicta curva punctum utcunque, veluti D, ductâque D E in angulo A E D, dato vel assumpto æquali, secante diametrum G B in B : erit, ex constructione, B E ∞ $\frac{bx}{2a}$; ideoque si E D vocetur y, erit D B ∞ $y - \frac{bx}{2a}$ seu z; F G ∞ $\frac{2ab}{c}$, G A ∞ $\frac{dde}{2ab}$; A B ∞ $\frac{ex}{2a}$, totaque G B ∞ $\frac{dde}{2ab} + \frac{ex}{2a}$. At cum ex proprietate Parabolæ D B quadratum sit æquale rectangulo F G B, erit, factâ multiplicatione ipsius z in se ipsam, atque $\frac{2ab}{c}$ in $\frac{dde}{2ab} + \frac{ex}{2a}$, $zz \infty dd + bx$. Vnde substituto $y - \frac{bx}{2a}$ loco z, obtinebitur $yy - \frac{bxy}{a} + \frac{bbxx}{4aa} \infty bx + dd$, id est, $yy - \frac{bxy}{a} \infty - \frac{bbxx}{4aa} + bx + dd$. Quod erat demonstrandum.

Quomodo autem pro casu hujus exempli converso Parabola describenda sit, ex comparatione ejusdem cum antedictis facile est colligere.

Si æquatio fuerit $\frac{bcy}{a} + by - \frac{bbyy}{aa} + \frac{1}{4}cc \infty xx + \frac{2byx}{a} - cx$,

[266]

and if we put according to the Rule

$$v = x + \frac{by}{a} - \frac{1}{2}c \text{, then } x = v - \frac{by}{a} + \frac{1}{2}c.$$

If we substitute this for x and its square for x^2, and delete all terms that cancel each other and arrange all terms as usual, then the above equation will have the following form

$$by + \frac{1}{2}c^2 = v^2.$$

From this it is clear that the equation has been reduced to the converse form of Theorem VIII and, therefore, that the required locus is a parabola. Its accurate determination results as it were of itself from what has already been explained. Therefore it will be sufficient to sketch it briefly by means of the added figure. In this figure we assume that AE (supposed to be indefinite) is the unknown quantity x and that this x includes an angle with the other unknown quantity y that is equal to the angle EAC or its supplement.

Determination of the locus

AE (supposed to be indefinite) equals x.

ED and all its parallel segments are equal to y.

$AK = \frac{1}{2}c = CH$ because KH is parallel to AC [2.18].

KH, or y, is to HB as a is to b. Therefore $HB = by/a$ and $DB = x - \frac{1}{2}c + (by/a) = v$.

KH, or y, is to KB as a is to e, so KB (on which the diameter is situated) equals ey/a. Division of by by ey/a yields ab/e; therefore the latus rectum FG is equal to ab/e [2.19]. The completely known term in the equation, namely $\frac{1}{2}c^2$, divided by

266   E L E M. C V R V A R V M

aſſumpto juxta Regulam $v \infty x + \frac{by}{a} - \frac{1}{2}c$, erit $x \infty v - \frac{by}{a} + \frac{1}{2}c$.

quo ſubſtituto in locum ipſius x, ejuſdemque quadrato loco xx, ablatisque iis, quæ ſe invicem deſtruunt, atque omnibus ritè ordinatis, æquatio ſuperior ſequenti formâ erit induta.

$$by + \tfrac{1}{2}cc \infty vv.$$

Vnde apparet eandem eſſe reductam ad formulam prædicti Theorematis V I I I converſim, ac proinde Locum quæſitum eſſe Parabolam. Cujus ſpecifica determinatio (ſuppoſitis, ut in adjuncta figura, A E indefinitè aſſumptam eſſe quantitatem incognitam x, atque cum altera y conſtituere angulum æqualem angulo E A C vel ejuſdem ad binos rectos ſupplemento) quoniam ex jam ante explicatis quaſi ſponte profluit, idcirco eam adjunctâ figurâ breviter indicaſſe ſuffecerit.

Determinatio Loci.

A E indefinitè ∞x.

E D omnesque ipſi parallelæ ∞y.

A K $\infty \frac{1}{2}c \infty$ C H, quia K H parallela A C.

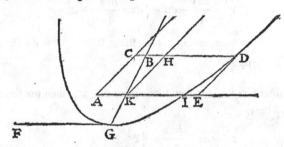

Vt a ad b, ita K H ſeu y ad H B: unde H B fit $\infty \frac{by}{a}$, & D B ∞x $- \frac{1}{2}c + \frac{by}{a} \infty v$.

Vt a ad e, ita K H ſeu y ad K B: unde K B (in qua diameter) fit $\infty \frac{ey}{a}$.

by diviſum per $\frac{ey}{a}$, reddit $\frac{ab}{e}$: unde latus rectum F G fit $\infty \frac{ab}{e}$.

$\frac{1}{2}cc$, nempe terminus æquationis in totum cognitus, diviſus per $\frac{ab}{e}$

[267]

ab/e, namely the latus rectum, yields $c^2e/2ab$, whence $KG = c^2e/2ab$ and $GB = (c^2e/2ab) + ey/a$.

Demonstration

The rectangle FGB is equal to the square on BD, so $\frac{1}{2}c^2 + by = v^2$ or

$$by = v^2 - \tfrac{1}{2}c^2 \text{, i.e.}$$

$$by = x^2 + \frac{2byx}{a} + \frac{b^2y^2}{a^2} - cx - \frac{bcy}{a} + \tfrac{1}{4}c^2 - \tfrac{1}{2}c^2.$$

Therefore, deleting what has to be deleted and after a proper rearrangement, the result is

$$\frac{bcy}{a} + by - \frac{b^2y^2}{a^2} + \tfrac{1}{4}c^2 = x^2 + \frac{2byx}{a} - cx .$$

Which was proposed.

Example of the reduction of equations to the form of Theorem IX

Let the equation be

$$y^2 + \frac{bxy}{a} - cy = ax - \frac{b^2x^2}{4a^2} - c^2 .$$

Let us put according to the Rule, $z = y + (bx/2a) - \tfrac{1}{2}c$, whence $y = z - (bx/2a) + \tfrac{1}{2}c$.

If we substitute this for y and its square for y^2, then the terms of the equation will become as follows

$$z^2 = ax - \frac{bcx}{2a} - \tfrac{3}{4}c^2 .$$

Let us, for convenience's sake, write d instead of $a - (bc/2a)$, (supposing a to be greater than $bc/2a$). Then the equation will be $z^2 = dx - (3c^2/4)$. And it is clear that this equation has been reduced to the form of Theorem IX and, therefore, that the required locus is a parabola. It will be very easy to determine and to describe this locus by means of what has already been explained [2.20] as may be gathered from the following figure and the brief remarks above it.

Determination of the locus

Let point A be the immutable initial point of x [2.21]. AE (supposed to be indefinite) equals x. ED and all its parallel segments equal y. EAK or AED is the angle that y and x have to enclose.

$\frac{a\,b}{c}$, nempe per latus rectum, reddit $\frac{cce}{2ab}$: undeK G fit ∞ $\frac{cce}{2ab}$, atque G B ∞ $\frac{cce}{2ab} + \frac{ey}{a}$.

Demonſtratio.

Rectangulum F G B ∞ B D quadrato, ergo $\frac{1}{2}cc + by \infty vv$, vel $by \infty vv - \frac{1}{2}cc$, hoc eſt, $by \infty xx + \frac{2byx}{a} + \frac{bbyy}{aa} - cx - \frac{bcy}{a} + \frac{1}{4}cc$.

$- \frac{1}{2}cc$.

Quocirca deletis delendis, factâque decenti tranſpoſitione, fiet $\frac{bcy}{a} + by - \frac{bbyy}{aa} + \frac{1}{4}cc \infty xx + \frac{2byx}{a} - cx$. Quod erat propoſitum.

Exemplum reductionis æquationum ad formulam Theorematis IX.

Sit æquatio $yy + \frac{bxy}{a} - cy \infty ax - \frac{bbxx}{4aa} - cc$. Aſſumatur juxta Regulam $z \infty y + \frac{bx}{2a} - \frac{1}{2}c$, eritque $y \infty z - \frac{bx}{2a} + \frac{1}{2}c$. Quo ſubſtituto in locum ipſius y, & ejus quadrato loco yy, fient æquationis termini, ut ſequitur: $zz \infty ax - \frac{bcx}{2a} - \frac{3}{4}cc$. Facilitatis ergo pro $a - \frac{bc}{2a}$ ſcribatur d, ſupponendo a eſſe majorem quàm $\frac{bc}{2a}$, eritque æquatio $zz \infty dx - \frac{3}{4}cc$. Et apparet eandem reductam eſſe ad formulam Theorematis IX, ac propterea Locum quæſitum eſſe Parabolam, quàm ex iis, quæ jam explicata ſunt, determinare ac deſcribere facillimum erit; ut ex ſequenti figura iisque quæ ſuper eâdem breviter annotata ſunt, colligere licebit.

Determinatio Loci.

Sit initium immutabile ipſius x punctum A.

A E indefinitè ∞ x.

E D omnesque ipſi parallelæ ∞ y.

E A K vel A E D, angulus quem x & y comprehendere debent.

[268]

$AK = \frac{1}{2}c$

KH is parallel to *AE*.

KH, or *x*, is to HB as 2*a* is to *b*, so *HB* = *bx*/2*a*.

KH, or *x*, is to *KB* as 2*a* is to *e*, so *KB* (on which the diameter is situated) equals *ex*/2*a*. Division of *dx* by *ex*/2*a* yields 2*ad*/*e*, so the latus rectum, which we call *FG*, will be equal to 2*ad*/e. Division of 3*c*²/4 by 2*ad*/e yields 3*c*²*e*/8*ad*, so *KG* becomes 3*c*²*e*/8*ad* and

$$GB = \frac{ex}{2a} - \frac{3c^2e}{8ad}.$$

Hence, if the parabola with vertex *G* has been described with *GB* as diameter and *FG* as latus rectum, and intersecting *KH* at *I*, then *ID* will be the required locus.

Demonstration

Let point *D* be selected at random on *ID* and let *DE* be drawn parallel to *AK*. If we put $DE = y$, then $HD = y - \frac{1}{2}c$ and $DB = y - \frac{1}{2}c + (bx/2a) = z$.

As its square equals the rectangle *FGB*, we have $z^2 = dx - \frac{3}{4}c^2$, that is

$$y^2 - cy + \frac{1}{4}c^2 + \frac{bxy}{a} - \frac{bcx}{2a} + \frac{b^2x^2}{4a^2} = ax - \frac{bcx}{2a} - \frac{3}{4}c^2.$$

Hence, if on either side equal terms are removed and the terms are arranged as usual, we will have

$$y^2 + \frac{bxy}{a} - cy = ax - \frac{b^2x^2}{4a^2} - c^2.$$

Which was proposed.

It will also not be difficult to reduce and to solve the counterpart [2.22] of this example as well as other related cases in a similar way by means of what we stated before.

268   E L E M. C V R V A R V M

$A K \infty \frac{1}{2} c.$

K H parallela ipfi A E.

Vt 2 a ad b, ita K H feu x ad H B : unde H B erit $\infty \frac{bx}{2a}$.

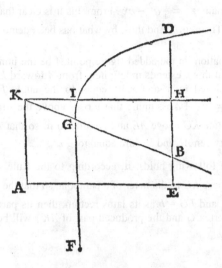

Vt 2 a ad e, ita K H feu x ad K B : unde K B (in quâ diameter) $\infty \frac{ex}{2a}$.

$d x$ divifum per $\frac{ex}{2a}$, reddit $\frac{2ad}{e}$: unde latus rectum, quod fit F G, erit $\infty \frac{2ad}{e}$.

$\frac{3}{4} c c$ divifum per $\frac{2ad}{e}$, reddit $\frac{3cce}{8ad}$: unde K G fit $\infty \frac{3cce}{8ad}$, atque G B $\infty \frac{ex}{2a} - \frac{3cce}{8ad}$.

Hinc fi G B diametro & latere recto F G per verticem G defcripta fit Parabola, fecans K H in I, erit I D Locus quæfitus.

Demonftratio.

Efto punctum D utcunque fumptum in I D , & D E ducta parallela ipfi A K, quæ fi vocetur y; erit H D $\infty y - \frac{1}{2} c$, ac D B $\infty y - \frac{1}{2} c + \frac{bx}{2a}$, hoc eft, z. Cujus quadratum cum æquetur rectangulo F G B, erit $z z \infty d x - \frac{3}{4} c c$, hoc eft, $y y - c y + \frac{1}{4} c c + \frac{bxy}{a} - \frac{bcx}{2a} + \frac{bbxx}{4aa} \infty a x - \frac{bcx}{2a} - \frac{3}{4} c c.$ Ac proinde, fi utrinque demantur æquales, terminique ritè tranfponantur, habebitur $y y + \frac{bxy}{a} - c y \infty a x - \frac{bbxx}{4aa} - c c.$ Quod erat propofitum.

Atque hujus quidem exempli converfum, ut & cæteros cafus huc fpectantes, ex iis, quæ jam dicta funt, fimili modo reducere atque refolvere non difficile erit.

Exem-

[269]

Examples of the reduction of equations to the form of Theorem X

If the equation is $ay - y^2 = bx$ or, what is the same, $y^2 - ay + bx = 0$ and if, according to the Rule, we put $z = y - \frac{1}{2}a$, then $y = z + \frac{1}{2}a$. If we substitute this for y and its square for y^2, then there will remain $z^2 = \frac{1}{4}a^2 - bx$. From this it is clear that the equation has been reduced to the case of Theorem X and thus, by what has been demonstrated there, the required locus is a parabola.

Let for its accurate determination, in the added figure point A be the immutable initial point of x and let us suppose that this x extends indefinitely from A toward E. Let, further, the given or chosen angle enclosed by y and x be equal to the angle EAK or to its supplement. Then, as we put $z = y - \frac{1}{2}a$ assuming that y rises above the straight line AE, we must also draw the straight line KG above AE and parallel to it, so that AK and all its parallel segments intercepted between AE and KG are equal to $\frac{1}{2}a$

After these preparations the following holds: if, according to the Rule, we make KG equal to $\frac{1}{4}(a^2 / b)$ and if we choose KG as diameter of a parabola, so that the ordinate-wise applied lines are parallel to AK, and $FG = b$ as its latus rectum, then its part described as GDI, intercepted between the vertex G and the produced part of AK, will be the required locus [2.23].

Lib. II. Cap. II.   269

Exempla reductionis æquationum ad formulam Theorematis X.

Si æquatio sit $ay - yy \infty\ b x$, sive, quod idem est, $yy - ay + b x \infty\ o$, assumpto juxta Regulam $z \infty y - \frac{1}{2} a$, erit $y \infty z + \frac{1}{2} a$. Quo substituto in locum ipsius y, & ejusdem quadrato loco yy, remanebit $zz \infty \frac{1}{4} aa - b x$. Vnde apparet, eandem esse reductam ad casum Theorematis X, ideoque per ea, quæ ibidem sunt demonstrata, Locum quæsitum esse Parabolam.

Ad cujus specificam determinationem esto in apposita figura ipsius x initium immutabile punctum A, eademque x se indefinitè ab A versùs E extendere intelligatur; sit autem datus vel assumptus angulus, quem y & x comprehendunt, æqualis angulo E A K,

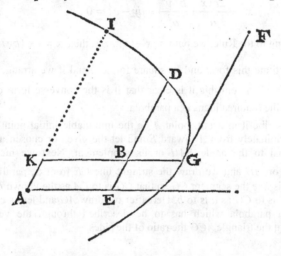

aut ipsius ad binos rectos complemento. Deinde, quoniam z assumpta est $\infty y - \frac{1}{2} a$, si y supra rectam A E exurgere intelligatur, ducenda quoque est supra ipsam recta K G ipsi A E parallela, ita ut A K omnesque ipsi æquidistantes inter A E & K G interceptæ sint $\infty \frac{1}{2} a$. Quo facto, si juxta Regulam fiat K G $\infty \frac{aa}{4b}$, eademque sumatur pro Parabolæ diametro, ad quam ordinatim applicatæ sint ipsi A K parallelæ, cujusque latus rectum F G sit ∞b: erit ipsius portio descripta G D I, quæ inter verticem G & productam A K intercipitur, Locus quæsitus.

Ll 3   Etenim

[270]

Indeed, if we select a point at random on the curve *GDI*, say *D*, and if we draw *DE* parallel to *AK*, intersecting the diameter *KG* at *B*, and if we put $DE = y$, then $DB = y - \frac{1}{2}a$ or z, and *GB*, or $GK - KB$, is equal to $\frac{1}{4}(a^2 / b) - x$.

As, by the nature of a parabola, the square on *BD* is equal to the rectangle *FGB*, it follows that

$$z^2 = \tfrac{1}{4}a^2 - bx, \text{ i.e } y^2 - ay = \tfrac{1}{4}a^2 - bx,$$

or $y^2 - ay + bx = 0$, or also $ay - y^2 = bx$. Which was proposed.

If the equation had been

$$\frac{b^2 y^2}{a^2} + dy - c^2 = \frac{2byx}{a} - x^2$$

or, which is the same,

$$x^2 - \frac{2byx}{a} + \frac{b^2 y^2}{a^2} + dy - c^2 = 0,$$

and if, according to the Rule, we put $v = x - (by / a)$, then $x = v + (by / a)$.

If we substitute this for *x* and its square for x^2, and if we arrange all terms as usual, then $c^2 - dy = v^2$. From this it is clear that it is the converse form of Theorem X and, therefore, that the required locus is a parabola.

Let, to describe it in detail, point *A* be the immutable initial point of *x* and let this *x* extend indeterminately from *A* toward *E* and let the given or chosen angle enclosed by *y* and *x* be equal to the angle *EAH* or its supplement. Next we measure the segment $AC = c^2 / d$ on *AH* and we draw the straight line *CF* from *C*, parallel to *AE*. Then we measure on this line the segment *CG* so that *CG* is to *CA* as the known *b* is to the known *a*, i.e. so that *AC* is to *CG* as *a* is to *b*. Hereafter we draw *AG* and let us choose this line as a diameter of a parabola which has to be described through the vertex *G* toward *A*. Furthermore, in the triangle *ACG* the ratio of the sides

270   E L E M. C V R V A R V M

Etenim aſſumpto in curva G D I punĉto utcunque, veluti D, ductâque D E ipſi A K parallelâ, quæ ſecet diametrum K G in B, ſi eadem D E vocetur y: erit D B ∞ $y - \frac{1}{2}a$ ſeu z, ac G B ſive G K — K B ∞ $\frac{aa}{4b} - x$. Hinc, cum ex natura Paraboles quadratum ex B D ſit æquale reĉtangulo F G B, erit $zz \infty \frac{1}{4}aa - bx$, hoc eſt, $yy - ay + \frac{1}{4}aa \infty \frac{1}{4}aa - bx$, ſive $yy - ay + bx \infty 0$, ſive etiam $ay - yy \infty bx$. Quod erat propoſitum.

Si æquatio fuerit $\frac{bbyy}{aa} + dy - cc \infty \frac{2byx}{a} - xx$, ſive, quod idem eſt, $xx - \frac{2byx}{a} + \frac{bbyy}{aa} + dy - cc \infty 0$: aſſumpto juxta Regulam $v \infty x - \frac{by}{a}$, erit $x \infty v + \frac{by}{a}$. Quo ſubſtituto in locum ipſius x, ejuſdemque quadrato loco xx, fiet, omnibus ritè ordinatis, $cc - dy \infty vv$. Vnde apparet caſum eſſe Theorematis X converſim, ac proinde Locum quæſitum eſſe Parabolam. Quæ quidem ut ſpecificè deſcribatur, eſto ipſius x initium immutabile A punĉtum, intelligaturque eadem x ſe extendere ab A verſùs E indeterminatè, ſitque angulus datus vel aſſumptus, quem y

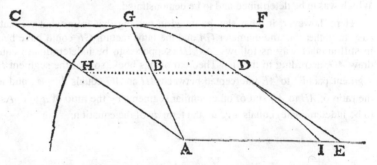

& x comprehendunt, æqualis angulo E A H aut ipſius ad duos reĉtos complemento. Deinde ſumatur in A H reĉta A C $\infty \frac{cc}{d}$, ducaturque ex C reĉta C F ipſi A E parallela, atque in eadem ſumptâ C G, quæ ſe habeat ad C A, ut cognita b ad a cognitam, hoc eſt, ut ſit uti a ad b, ita A C ad C G, agatur A G, eaque pro diametro Parabolæ ſumatur, quæ per verticem G versùs A erit deſcribenda. Porrò cum in triangulo A C G ob rationem cognitam laterum
A C,

[271]

AC and *CG* enclosing the known angle *C*, is known; the latter is namely equal to the given or chosen angle or to its supplement; therefore the ratio of *AC* to *AG* is also known, let this be as *a* to *e*. So, as *AC* appears to be c^2 / d, therefore $AG = c^2 e / ed$. If the wholly known term of the equation, namely c^2, is divided by this *AG*, then ad / e will result as the latus rectum [2.24]. Hence if $GF = ad / e$, then *GF* will be the latus rectum of the required parabola corresponding to the diameter *GA*. So, if we describe a parabola, just as *GDI*, on the said diameter and with the said latus rectum, intersecting *AE* at *I*, then, I say: the curve *IDG* is the required locus.

Indeed, if we select a point at random on this curve, say *D*, and if we draw *DE* parallel to *AH*, and *DBH* parallel to *AE*, and if we represent *DE* by y, then also $AH = y$. As *AH* is to *HB* as *AC* is to *AG* (that is as *a* is to *b*), $HB = by / a$, and so $DB = x - (by / a)$ or *v*, as *DH* or *AE* is equal to *x*. In the same way the following holds: because *AH* or *y* is to *AB* as *AC* is to *AG* (that is as *a* is to *e*), therefore $AB = ey / a$ and $GA - AB$ (or *GB*) = $(c^2 e / ad) - (ey / a)$. By the nature of a parabola the rectangle *FGB* is equal to the square on *DB*, so, if we multiply *FG* or *ad/e* by *GB* (or $(c^2 e / ad) - (ey / a))$, and *BD*, or *v*, by itself, then $c^2 - dy = v^2$. This means, if we substitute again $x - (by / a)$ for *v*, that

$$c^2 - dy = x^2 - 2\frac{byx}{a} + \frac{b^2 y^2}{a^2} \quad \text{or} \quad \frac{b^2 y^2}{a^2} + dy - c^2 = 2\frac{byx}{a} - x^2$$

Which was to be determined and to be demonstrated.

Here, however, it must also be noted in passing that from what we stated before it is easy to gather that the diameter *GA* and the latus rectum *GF* could have been investigated in still another way as follows. As *AH* (supposed to be indeterminate) equals *y*, we first draw *AG*, according to the first Theorem of this book, so that the segment *HB*, just as any segment parallel to *AE* intercepted between *AH* an *AG* equals by / a, and let us represent the ratio of *AH* to *AB* and of other similar segments by the ratio of *a* to *e*. As *AB* (supposed to be indeterminate) equals ey / a, the term *dy* of the equation,

A C, C G, cognitum angulum C comprehendentium, utpote dato vel affumpto aut ejufdem ad duos rectos fupplemento æqualem, cognita item fit ratio, quam habet A C ad A G, quæ fit ut a ad e; erit, A C exiftente $\infty \frac{cc}{d}$, A G $\infty \frac{cce}{ad}$. Per quam fi terminus æquationis, in totum cognitus, nimirum cc, dividatur, orietur $\frac{ad}{e}$ pro latere recto. Ac proinde fi fiat G F $\infty \frac{ad}{e}$, erit G F latus rectum quæfitæ Parabolæ, diametro G A correfpondens; atque iccirco fi ad dictam diametrum, dictumque latus rectum Parabola defcribatur, ut G D I, fecans A E in I : dico I D G curvam effe Locum quæfitum.

Sumpto enim in ea puncto utcunque, veluti D, ductisque D E ipfi A H, ac D B H ipfi A E parallelis, fi eadem D E exprimatur per y, erit quoque A H ∞y. Cumque fit ut A C ad C G, id eft, ut a ad b, ita A H ad H B : erit H B $\infty \frac{by}{a}$, ideoque cum D H feu A E fit ∞x, erit D B $\infty x - \frac{by}{a}$ feu v. Similiter cum fit ut A C ad A G, hoc eft, ut a ad e, ita A H feu y ad A B : erit A B $\infty \frac{ey}{a}$, & G A — A B feu G B $\infty \frac{cce}{ad} - \frac{ey}{a}$. Hinc cum ex natura Paraboles rectangulum F G B fit æquale quadrato ex B D, erit, factâ multiplicatione ipfius F G feu $\frac{ad}{e}$ in G B feu $\frac{cce}{ad} - \frac{ey}{a}$, & ipfius B D feu v in fe ipfam, $cc - dy \infty vv$. Hoc eft, reftituto $x - \frac{by}{a}$ loco v, erit $cc - dy \infty xx - \frac{2byx}{a} + \frac{bbyy}{aa}$, vel $\frac{bbyy}{aa} + dy - cc \infty \frac{2byx}{a} - xx$.

Quod determinandum, demonftrandumque erat.

Obiter autem & hîc notandum, ut ex antedictis quoque facile eft colligere, aliter etiam diametrum G A atque latus rectum G F indagari potuiffe, hoc modo :

Cum A H indeterminatè fit ∞y, juxta primum Theorema hujus ita ducatur A G, ut recta H B, quemadmodum & quælibet alia ipfi A E parallela, quæ inter A H & A G intercipitur, fit $\infty \frac{by}{a}$; ponaturque ratio, quæ eft inter A H & A B fimilesque, ut a ad e : ideoque cum A B indeterminatè fit $\infty \frac{ey}{a}$, terminus æquationis dy

per

[272]

divided by ey/a, will show that the latus rectum FG of the conic section equals ad/e. In the same way the term c^2 of the equation divided by the said latus rectum or ad/e, will give the quotient c^2e/ad as the required AG [2.25].

It would be superfluous to add more examples here, as we intend to explain and to describe all possible cases by a general rule hereafter [2.26].

Moreover, although we did not illustrate the Rules explained in the first chapter, there by particular examples or by hypothetical cases, we do not even think it in any way necessary to do this here or in what follows. Surely, everyone who has understood the Rules themselves in the right way, is able to apply these easily to any proposed example or hypothetical case. However, since in the first book we have purposely omitted some important properties of the parabola, hyperbola and ellipse, considering that they could suitably be proposed and demonstrated by way of problems in the right places in this book and that at the same time they could be considered as particular examples of the proposed Rules, we will therefore add their explanation here and at the end of the following chapter.

PROBLEM I

Proposition 11

A point and a straight line being given, it is required to find another point in the plane passing through both, with the property that the two segments drawn from this point, one to the given point and the other perpendicular to the given line, are mutually equal and, because there are infinitely many points of this kind, to determine and to describe the locus on which they are all and each separately can be found.

Let A be the given point and BC the straight line whose position is given and let it be required to find another point in the plane passing through both,

per eandem divifus oftendet latus rectum fectionis F G ∞ $\frac{ad}{e}$. Similiter terminus æquationis cc per prædictum latus rectum feu
$\frac{ad}{e}$ divifus dabit quotientem $\frac{cce}{ad}$ pro quæfita A G.

Plura hîc exempla fubjungere fupervacuum foret, cum mox
omnes omnino cafus poffibiles generali regulâ annotare ac de-
monftrare animus fit.

Porrò quamvis Regulas capite primo explicatas par-
ticularibus ibidem exemplis feu cafibus in hypothefi
non illuftraverimus, neque etiam id aut hîc aut in fe-
quentibus ullo modo neceffarium ducamus, quippe
cum unufquifque, qui Regulas ipfas rectè perceperit,
eafdem quibuflibet propofitis exemplis feu cafibus in
hypothefi facilè applicare valeat: quandoquidem tamen libro primo infignes quafdam proprietates Parabolæ, Hyperbolæ, atque Ellipfis confultò prætermifi-
mus, eâ mente, ut in hoc libro fuis locis per modum
Problematum non incongruè proponi ac demonftrari,
fimulque tanquam propofitarum Regularum particu-
laria exempla haberi poffent, earundem explicationem
hîc & fub finem fequentis capitis fubjiciemus.

P R O B L E M A I.

Propofitio 11.

Datis puncto & lineâ rectâ, in plano per utrumque
ducto aliud punctum invenire, à quo binæ rectæ, altera ad datum punctum, altera ad datam lineam perpen-
diculariter ductæ, fibi invicem fint æquales: & quoniam infinita funt ejufmodi puncta, quæ quæftioni fa-
tisfaciunt, Locum determinare ac defcribere, in quo
cuncta & fingula reperiantur.

Sit datum punctum A, & data pofitione recta linea B C, oporteatque in plano quod per utrumque ducitur, aliud punctum invenire,

[273]

say D, so that the segments DA and DF are mutually equal; the lattter being supposed to be perpendicular to the given BC. First we draw the perpendicular AE, which we will call a. Let according to the Rule the two unknown and indeterminate segments EF and FD enclosing the given right angle EFD be supposed to be known and determinate and let the former, EF, be called x and the latter, FD, be called y. If we suppose that, moreover, AG has been drawn parallel to EF, then in the right-angled triangle AGD the base AD equals y as being equal to the drawn DF; the side AG or the segment EF, however, equals x and $GD = | y - a |$, as GD is either $FD - AE$ (if point G falls between D and E) or $AE - FD$ (if point D falls between F and G) [2.27]. Since the square on the base equals the two squares on the sides, taken together, therefore the equation will be

$$y^2 = x^2 + y^2 - 2ay + a^2,$$

that is

$$2ay - a^2 = x^2,$$

if we omit the terms that cancel each other mutually and if we arrange all terms as usual. This, however, is the converse form of Theorem IX of this book and so the required locus will be a parabola. If, according to what

nire, quemadmodum D; ita ut ductæ rectæ D A, D F, quarum
hæc ad datam B C intelligitur perpendicularis, sibi invicem æquales sint.

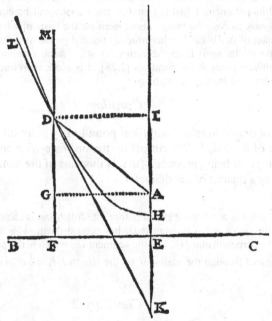

Ductâ perpendiculari A E, quæ vocetur *a*, ac suppositis juxta
Regulam binis lineis E F, F D incognitis atque indeterminatis
datum angulum rectum E F D comprehendentibus tanquam cognitis ac determinatis, quarum prior E F vocetur *x*, ac posterior
F D nominetur *y*; si ductâ præterea intelligatur A G ipsi E F
æquidistans, erit in triangulo rectangulo A G D basis A D ∞ *y*,
utpote ∞ ductæ D F; latus verò A G seu recta E F ∞ *x*, & G D,
sive (si punctum G cadat inter D & F) F D — A E, aut (si punctum D inter F & G cadat) A E — F D ∞ *y* = *a*. Vnde, cum quadratum basis æquale sit binis laterum quadratis simul sumptis, æ-
quatio erit *y y* ∞ *x x* + *y y* — 2 *a y* + *a a*, hoc est, ablatis iis quæ se
invicem destruunt, omnibusque ritè ordinatis, erit 2 *a y* — *a a* ∞ *x x*.
Qui quidem casus est Theorematis noni hujus libri conversim,
ac proinde Locus quæsitus erit linea Parabolica. Quare si juxta

Pars II.   Mm   ea,

[274]

has been explained there, we draw the straight line *EI* extended indefinitely and parallel to *FD*, and if we subtract the segment *EH* from it equal to $a^2 / 2a$, i.e. $\frac{1}{2}a$, then the diameter of the parabola that has to be described will be situated on the said *EI* (which diameter, however, is also the axis because the angle *EFD* is a right angle), its vertex, however, will be at *H* and its parameter (= latus transversum, transl.) will be equal to $2a$. Therefore it will be very easy to describe the parabola itself on the basis of what has been explained in the first chapter of this book. Furthermore, as point *A* on the axis, (which is namely a fourth part of the parameter apart from the vertex *H*) is the same as what is generally named the focus or umbilical point of the parabola [2.28], it is clear, therefore, that what follows can correctly be inferred from the foregoing.

Corollary 1

The segment drawn from the umbilical point to any point on the parabola, is equal to that part of the axis [2.29] cut off by the ordinate-wise applied line that passes through this point being produced (that is this part of the diameter, transl.) through the vertex by a quarter of the diameter.

Indeed, on the basis of what we stated before, the following is clear: if we select point *D* at random on the curve and if through it *DI* has been ordinate-wise applied to the axis, then *AD* equals the perpendicular *DF*, i. e. the segment *IE*, which is the part of the axis cut off by *DI* and produced through the vertex *H* by the length $HE = \frac{1}{2}a$, i.e. the fourth part of the parameter.

Corollary 2

On the basis of what we stated before, the following is evident: if we suppose the same as above and if we produce *FD*, e.g. up to *M*, and if we draw the tangent at the selected point *D*, e.g. *LDK*, then the angle *FDK* or *MDI* is equal to the angle *ADK*.

Indeed, let the tangent *LDK* meet the produced part of the axis at *K*; then[1] the segment *IH* equals *HK* [2.30] and so (if we add on both sides the mutually equal *HE* and *AH*), the segment *IE*, i.e. *AD* equals *AK* and therefore[2] it is also necessary that angle *ADK* equals angle *AKD*, i.e. angle *FDK* or *MDL* [2.31].

---

[1] Lib. I, Prop. 2, Cor. 2, p. [174] (the text erroneously gives Cor. 1).

[2] I, 5.

274   E L E M. C V R V A R V M

ea, quæ ibidem expoſita ſunt, ex E ducatur recta E I indefinitè extenſa atque ipſi F D æquidiſtans ; & ab eadem auferatur recta
E H $\infty \frac{a\,a}{2\,a}$, id eſt, $\frac{1}{2}a$: erit deſcribendæ Parabolæ diameter in dicta E I, (quæ quidem diameter axis quoque eſt, propter angulum
E F D rectum) vertex autem in H, ac parameter ∞ 2 a. Vnde,
per ea quæ libri primi capite primo expoſita ſunt, Parabolam
ipſam deſcribere facillimum erit. Cumque porrò axis punctum A,
utpote quod ab H vertice diſtat quartâ ipſius parametri parte, id
ipſum ſit, quod vulgò Parabolæ Focus ſeu Vmbilicus nuncupa-
tur, apparet ex præmiſſis rectè inferri, quæ ſequuntur.

Corollarium 1.

Quæ ab Vmbilico ad quodlibet Parabolæ punctum
recta ducitur æqualis eſt axis portioni per applicatam
ab eodem puncto abſciſſæ & quadrante parametri per
verticem productæ.

Conſtat enim ex antedictis rectam A D, utcunque aſſumptum
fuerit in curva punctum D, ſi per idem illud ad axem ordinatim
applicata ſit D I, æqualem eſſe perpendiculari D F, hoc eſt, rectæ
I E, nempe axis portioni, per applicatam D I abſciſſæ, & per
verticem H, longitudine H E $\infty \frac{1}{2}a$, id eſt, quadrante parametri,
productæ.

Corollarium 2.

Manifeſtum quoque eſt ex antedictis, ſi poſitis quæ
ſupra, & productâ F D, uti ad M, per aſſumptum punctum D contingens ducta ſit, ut L D K, angulum F D K
ſive M D L angulo A D K æqualem eſſe.

[1] per 1.
Cor. 2
primi hu-
jus.
[2] per 5
primi.

Occurrat enim contingens L D K axi producto in K, eritque [1]
recta I H ipſi H K, ideoque (æqualibus H E, A H utrinque additis) recta I E, hoc eſt, A D, ipſi A K æqualis ; ac proinde [2] & angulus A D K angulo A K D, hoc eſt, angulo F D K ſive M D L æ-
qualis ſit neceſſe eſt.

C A-

[275]

CHAPTER III

I N the third case, however, formulated previously [3.1], in which namely both unknown quantities are raised to the square or one of them is found in the equation, multiplied by the other and the equation cannot be reduced to a simpler form, then we will arrive at one of the following forms:

I. $$yx = f^2.$$

II. $$\frac{ly^2}{g} = x^2 - f^2.$$

III. $$y^2 - f^2 = \frac{lx^2}{g}.$$

IV. $$\frac{ly^2}{g} = f^2 - x^2.$$ [3.2]

THEOREM XI

Proposition 12

If the equation is $yx = f^2$, then the required locus is a hyperbola

Indeed, as in the foregoing, let point A be the immutable initial point of x and let us suppose that this x extends indefinitely along the straight line AE and let the given or chosen angle, enclosed by y and x be equal to the angle EAB or to its supplement. Next we measure the segment AC (equal to f) along AE and we draw CG equal to AC and parallel to AB and we describe[1] the hyperbola GD through point G and

---

[1] On the basis of what has been proposed in the Corollary of Prop. 11 and 12 (Lib. I, p. [101]) and in the last chapter of Liber Primus.

L I B. II. C A P. III.   275

C A P V T III.

TErtio autem cafu fupra expreſſo, cùm nempe quantitatum incognitarum utraque ad quadratum aſcendit, five altera in alteram ducta in æquatione reperitur, neque æquatio ad terminos magis fimplices reduci poteſt, ad aliquam fequentium formularum deventum erit;

$$\text{I. } yx \infty ff.$$

$$\text{II. } \frac{lyy}{g} \infty xx - ff.$$

$$\text{III. } yy - ff \infty \frac{lxx}{g}.$$

$$\text{IV. } \frac{lyy}{g} \infty ff - xx.$$

T H E O R E M A XI.
Propofitio 12.

Si æquatio fit $yx \infty ff$, Locus quæfitus eſt Hyperbola.

Sit enim, ut in præcedentibus, ipſius x initium immutabile A

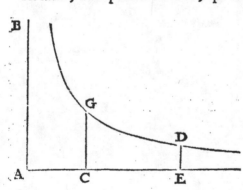

punctum, atque ea-
dem illa x per rectam
A E indefinitè fe ex-
tendere intelligatur;
fitque datus vel aſſum-
ptus angulus, quem y
& x comprehendunt,
æqualis angulo E A B,
aut ejufdem ad binos
rectos fupplemento.
Deinde fumatur in
A E recta A C ∞f, ducaturque C G eidem
æqualis ac ipfi A B parallela, defcriptâque [1] per punctum G at- [1] *per ea-*
que in
M m 2   que *Corol. ad*
11 *&* 12, *nec non cap. ult. lib. primi hujus tradita fûnt.*

[276]

having AE and AB as its asymptotes. Then, I say: the curve GD is the required locus.

Indeed, if we select a point at random on this curve, e.g. D, and if we draw DE parallel to AB, then, by the nature of a hyperbola[1], the rectangle AED will be equal to the rectangle ACG, i.e. to the square on AC. So, if we assume AE as the unknown quantity x and if ED is called y, then $yx = f^2$. Which was to be determined and demonstrated.

THEOREM XII

Proposition 13

If the equation is $ly^2 / g = x^2 - f^2$, then the required locus will be a hyperbola.

Indeed, l is either equal to g or unequal and if l equals g, then the above equation would be the same as $y^2 = x^2 - f^2$ (let it be sufficient to point this out once) The following is easily understood: if point A is the immutable initial point of x and if we suppose that this x extends indefinitely from A toward E along the line AE and if the given or chosen angle enclosed by y and x is equal to the angle AGF and if both AG and AC are made equal to the known f, and if GF is chosen equal to GC, and if a hyperbola, say GD, is described[2] with A as its center, with GC as its transverse diameter (equal to its latus rectum or parameter GF), then this curve GD will be the required locus. [3.3]

Indeed, if we select a point at random on it, say D, and if we draw DE parallel to FG, then,[3] by the nature of a hyperbola, and because we suppose CG and GF to be equal, the square on DE will be equal to the rectangle CEG.

---

[1] Lib. I, Prop. 3, p. [180], [181].

[2] On the basis of what has been shown in the last chapter of Liber Primus.

[3] Lib. I, Prop. 10, p. [196].

276   E L E M. C V R V A R V M

que Aſymptotis A E, A B Hyperbolâ G D : dico curvam G D
eſſe Locum quæſitum.

Sumatur enim in eadem curva punctum utcunque, veluti D,
ductâque D E ipſi A B parallelâ, erit ex natura Hyperboles [1] rectangulum A E D rectangulo A C G, hoc eſt, quadrato ex A C
æquale. Hinc, cum A E ſit aſſumpta pro incognita quantitate *x*,
ſi E D vocetur *y*, erit *y x* ∞ *ff*. Quod determinandum, demonſtrandumque erat.

¹ per 3
primi hu-
jus.

T H E O R E M A   XII.

Propoſitio 13.

Si æquatio ſit $\dfrac{l y y}{g}$ ∞ *x x* —*ff*, erit Locus quæſitus linea Hyperbolica.

Aut enim *l* ipſi *g* æqualis eſt aut inæqualis, & ſi æqualis ſit, erit
ſuperior æquatio eadem ac ſi eſſet *y y* ∞ *x x* —*ff* (quod ſemel monuiſſe ſufficiat).Ac
facilè apparet, ſi i-
pſius *x* initium im-
mutabile ſit pun-
ctum A, atque ea-
dem *x* ſe in linea
A E ab A versùs E
indefinitè extende-
re intelligatur, ſit-
que angulus da-
tus vel aſſumptus,
quem *y* & *x* com-
prehendunt,æqua-
lis angulo A G F,

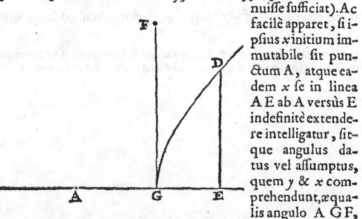

quòd ſi tam A G quàm A C fiant ∞ *f* cognitæ, ac G F ſumatur
∞ G C, centroque A, & transversâ diametro C G ipſi G F late-
ri recto ſive parametro æquali deſcribatur [2] Hyperbola, ut G D,
eandem curvam G D fore Locum quæſitum.

Sumpto enim in ea puncto utcunque, veluti D, ductâque D E
ipſi F G parallelâ, erit [3] ex natura Hyperboles, cùm C G & G F
ſupponantur æquales, quadratum ex D E æquale rectangulo
C E G.

² per ea
quæ cap.
ult. primi
hujus
oſtenſa
ſunt.
3 per 10
primi hu-
jus.

[277]

Hence (if *DE* is called *y*) we have $y^2 = x^2 - f^2$ because, by hypothesis, *CE* or *AE + AC* equals $x + f$ and *GE* or *AE − AG* equals $x − f$.

But, in fact, if *l* and *g* are unequal it is clear that $x^2 - f^2$ is to y^2 as *l* is to *g*. If then, according to what has been explained before, the parameter *GF* no longer equals the transverse diameter *CG*, but if the transverse diameter *CG* is to the parameter GF as *l* is to *g* and if all the rest is as above, then the question will be answered in the same way.

Indeed,[1] by the nature of a hyperbola, the square on *ED* is to the rectangle *CEG* as *FG* is to *GC*, that is, y^2 is to $x^2 - f^2$ as *g* is to *l*; so, if we reduce the proportion to a multiplication, $ly^2 = gx^2 - gf^2$. Therefore, if we divide both members of the equation by *g*, we will have $ly^2 / g = x^2 - f^2$. Which was to be determined and demonstrated.

THEOREM XIII

Proposition 14

If the equation is $y^2 - f^2 = lx^2 / g$, then the required locus will be a hyperbola [3.4].

Let, for its accurate determination, in the added figure point *A* be the immutable initial point of *x* and let us suppose that this *x* extends indefinitely from *A* toward *E*

---

[1] Lib. I, Prop. 10, p. [196].

L I B II. C A P III.   277

C E G. Hinc, ſi D E vocetur *y*, cum ex hypotheſi C E ſeu A E +A C ſit ∞ *x* +*f*, & G E ſive A E —A G ∞ *x* —*f*, erit *yy* ∞ *xx* —*ff*.

At verò ſi *l* & *g* ſint inæquales, apparet eſſe, ut *l* ad *g*, ita *x x* —*ff* ad *yy*. Ac proinde ſi juxta ea, quæ ſupra expoſita ſunt, non jam parameter G F diametro tranſverſæ C G æqualis, ſed ut *l* ad *g*,

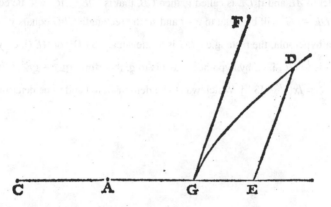

ita fiat tranſverſa diameter C G ad G F parametrum, cæteraque omnia, ut ſupra, eodem modo quæſito erit ſatisfactum.

Eſt enim [1] ex natura Hyperboles, ut F G ad G C, ita E D [1] *per 10 primi hujus.* quadratum ad C E G rectangulum, hoc eſt, ut *g* ad *l*, ita *y y* ad *x x* —*ff*, unde, revocando proportionem ad æqualitatem, erit *l yy* ∞ *g x x* —*g ff*. Ac proinde ſi utraque hujus æqualitatis pars dividatur per *g*, erit $\frac{l yy}{g}$ ∞ *x x* — *ff*. Quod determinandum, demonſtrandumque erat.

T H E O R E M A XIII.

Propoſitio 14.

Si æquatio ſit *yy* — *ff* ∞ $\frac{l x x}{g}$, erit Locus quæſitus Hyperbola.

Ad cujus determinationem ſpecificam eſto in appoſita figura ipſius *x* initium immutabile punctum A, ipſaque *x* ſe ab A versùs

[278]

along line AE and let the angle enclosed by y and x, be equal to the angle EAG or its supplement. Because $y^2 - f^2$ is to x^2 as l is to g, the following is immediately clear: if both AG and AC are supposed to be equal to the known f and if we make CG to GF as l to g (GF being parallel to AE) and if we next describe the hyperbola GD with A as its center, CG as its transverse diameter and GF as its parameter, then this curve GD will be the required locus.

Indeed, if we select a point at random on it, e. g. D, and if we draw DE parallel to AG and DB parallel to AE and if DE is called y, then CB, that is, $DE + AC$, will be equal to $y + f$, and BG, or $DE - AG$, will be equal to $y - f$ and so the rectangle CBG equals $y^2 - f^2$. By the nature of a hyperbola, the rectangle CBG is to the square on DB or AE (i. e. $y^2 - f^2$ is to x^2), as CG is to GF, i. e., by hypothesis, as l is to g, therefore $gy^2 - gf^2 = lx^2$, which means $y^2 - f^2 = lx^2 / g$ [3.3]. Which was to be demonstrated and to be determined.

278   E L E M. C V R V A R V M

E in linea A E indefinitè extendere intelligatur, sitque angulus
quem *y* & *x* comprehendunt æqualis angulo E A G aut ejusdem
ad duos rectos supplemento. Deinde, cum sit ut *l* ad *g*, ita *yy* — *ff*

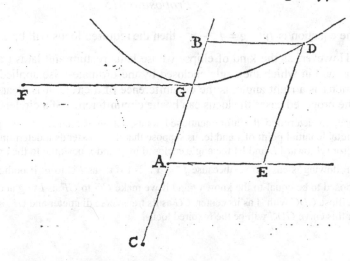

ad *xx*, statim apparet, si tam A G quam A C sumantur æquales
f cognitæ, fiatque ut *l* ad *g*, ita C G ad G F (quæ quidem G F sit
ipsi A E parallela), ac postea centro A, transversâ diametro C G,
& parametro G F Hyperbola describatur G D, eandem curvam
G D fore Locum quæsitum.

Sumpto namque in ea puncto utcunque, veluti D, ductâque
D E ipsi A G, ac D B ipsi A E parallelâ, si eadem D E vocetur *y*,
erit C B, hoc est, D E + A C, ∞ *y* + *f*; & B G, sive D E — A G,
∞ *y* — *f*, ideoque C B G rectangulum ∞ *yy* — *ff*. Dein cum [1] ex
natura Hyperbolæ sit ut C G ad G F, hoc est, ex hypothesi ut *l* ad
g, ita rectangulum C B G ad D B sive A E quadratum, id est, ita
yy — *ff* ad *xx* : erit *gyy* — *gff* ∞ *lxx*, hoc est, *yy* — *ff* ∞ $\dfrac{lxx}{g}$.

Quod demonstrandum, determinandumque erat.

[1] *per 10 primi hujus.*

THEO-

[279]

THEOREM XIV

Proposition 15

If the equation is $ly^2 / g = f^2 - x^2$, then the required locus will be an ellipse [3.5].

However, as the kind of ellipse whose latus rectum and latus transversum are equal and in which the angle enclosed by the ordinate-wise applied lines and the diameter is a right angle, is the circumference of a circle, it is clear therefore that in the proposed cases the locus can be the circumference of a circle.

Hence, to determine the aforementioned locus, let in the added figure point A be the immutable initial point of x and let us suppose that this x extends indeterminately along line AE from A toward E and let the angle enclosed by y and x be equal to the angle AGF. Then the following is easy to see because $f^2 - x^2$ is to y^2 as l is to g: if both AG and AC are supposed to be equal to the known f and if we make CG to GF as l to g and if we describe the ellipse GDC with A as its center, CG as its transverse diameter and GF[1] as its parameter, then this curve GDC will be the required locus.

---

[1] Lib. I, Prop. 13, Cor. 7, p. [213] and Lib. I, Prop. 14, Cor. 1, p. [218] and on the basis of what has been shown at the end of chapter IV of Liber I.

L i b. II. C a p. III.

T h e o r e m a XIV.

Propositio 15.

Si æquatio sit $\frac{lyy}{g}$ ∞ $ff - xx$, erit Locus quæsitus Ellipsis.

At verò cum Ellipseos species, quæ latera rectum & transversum æqualia habet, angulumque quem ordinatim applicatæ faciunt ad diametrum rectum, sit Circuli circumferentia : palam fit casu proposito Locum quæsitum etiam Circuli peripheriam esse posse.

Hinc ad prædicti Loci determinationem esto in apposita figura ipsius *x* initium immutabile A punctum, atque eadem *x* se per lineam A E ab A versùs E indeterminatè extendere intelliga-

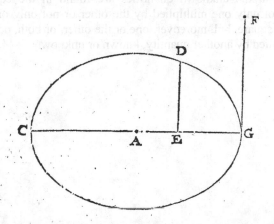

tur, sitque angulus, quem *y* & *x* comprehendunt, æqualis angulo A G F. Porrò cum sit ut *l* ad *g*, ita $ff - xx$ ad yy : facilè apparet, si tam A G quàm A C sumantur æquales *f* cognitæ; fiatque ut *l* ad *g*, ita C G ad G F, ac centro A, transversâ diametro C G, & parametro G F ' Ellipsis describatur G D C ' eandem curvam G D C fore Locum quæsitum.

per 7
Corol. 1 3,
& 1 Cor.
Sum- *14 primi*
hujus, ut & per ea quæ circa finem cap. 4 *ejusdem lib. tradita sunt.*

[280]

For if we select a point at random on it, say *D,* and if we draw *DE* parallel to *FG,* then,[1] by the nature of an ellipse [3.5], the square on *ED* will be to the rectangle *CEG* as *FG* is to *GC.* This means the following: if *ED* is called *y,* then (because $CE = f + x$ and $EG = f - x$),

y^2 will be to $f^2 - x^2$ as *g* is to *l,* whence $ly^2 / g = f^2 - x^2$. Which was proposed.

For the rest it is clear and certain that if *CG* and *GF* are equal, i. e. if $l = g$, the rectangle *CEG* will also be equal to the square on *ED.* Therefore the curve *GDC* will be the circumference of a circle if the angle *CGF* is a right angle.

General Rule and method of reducing all equations that result from a suitable operation (when the locus is a hyperbola or an ellipse or the circumference of a circle) to one of the preceding four cases that have already been explained by means of as many theorems

It may happen that the unknown quantities are found in the equation in the following way: not only one multiplied by the other or not only one of them or both raised to the square, but, moreover, one or the other, or both, occurring to the first power multiplied by another quantity, known or unknown

---

[1] Lib. I, Prop. 13, p. [205].

280   E L E M. C V R V A R V M

Sumpto namque in ea puncto utcunque, veluti D, ductâque

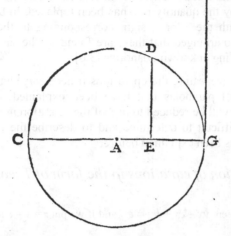

DE ipfi F G paral-
lelâ, erit [1] ex natu-
ra Ellipfeos ut F G
ad G C, ita E D qua-
dratum ad C E G re-
ctangulum. Hoc eft,
fi E D vocetur *y*, cum
C E fit ∞ *f + x*, &
E G ∞ *f — x*, erit ut
g ad *l*, ita *y y* ad *f f*
— *x x*, unde $\frac{lyy}{g}$ ∞ *f f*
— *x x*. Quod erat pro-
pofitum.

[1] per 13 primi hu-jus.

Cæterùm liquidò
conftat, fi C G & G F
æquales fuerint, hoc
eft, fi *l* ∞ *g*, quòd
etiam C E G rectan-
gulum quadrato E D
æquale fit futurum. Ideoque fi angulus C G F fit rectus, curvam
G D C fore Circuli circumferentiam.

*Regula univerfalis, modufque reducendi
omnes æquationes, quæ ex convenienti operatione exiftunt, cùm Locus vel Hyperbola eft, vel Ellipfis, vel Circuli circumferentia, ad aliquem quatuor cafuum præcedentium, totidem Theorematibus jam
explicatorum.*

Si contingat, ut quantitatum incognitarum non modò una in alteram, aut non tantùm alterutra vel utraque in fe ducta, fed & vel hæc, vel illa, vel utraque u-
nius præterea dimenfionis in æquatione reperiatur,
conftituens planum cum alia, five cognitâ five incognitâ,

[281]

or even multiplied by another quantity partially known, partially unknown. (e.g. ay/b, transl.). In this case it is necessary to replace the unknown quantities or one of them with one or two others that are greater or smaller than those. The difference is the entire quantity that has been multiplied by the quantity that has not been replaced, in the case namely in which none of the unknown quantities has been squared in the equation. But otherwise the difference is only half the quantity that has been multiplied by the quantity that has been replaced. In both cases the difference is provided with the $+$ or $-$ sign, corresponding to the sign of the quantities in question, if so arranged that these are found on the same side of the equation as the corresponding unknown quantities [3.6].

If one acts in this way (repeating the operations if necessary) and if one does not arrive at the forms of parabolas that have been explained in the second chapter, then the equation will be reduced to one of the four aforementioned cases and then it will not be difficult to determine and to describe the corresponding locus by means of what has been explained before.

Example of the reduction of equations to the form of Theorem XI

If the equation would have been $yx - cx + hy = e^2$ and if we put $z = y - c$ and $v = x + h$, then $z + c = y$ and $v - h = x$.

Hence, if according to the Rule, we substitute $z + c$ for y everywhere in the equation, then $zx + cx - cx + hz + hc = e^2$, or $zx + hz + hc = e^2$, and if we substitute once more $v - h$ for x, $zv - hz + hz + hc = e^2$, whence $zv = e^2 - hc$, or $zv = f^2$ (if we write f^2 instead of the term $e^2 - hc$, which is totally known). Then it is clear that the equation has been reduced to the form of Theorem XI and, therefore, that the required locus is a hyperbola [3.7].

Let, for its accurate determination and description [3.8], in the added figure point A be the immutable initial point of x and let us suppose that this x extends indefinitely along the straight line AE and let the angle enclosed by y and x be equal to the angle EAK

tâ, five etiam cum partim cognita & partim incognita
quantitate : oportet loco incognitarum , aut illarum
alterutrius, affumere alias vel aliam, quæ ipfas exce-
dunt , vel ab iis deficiunt ; idque integrâ quantitate,
quæ cum illa incognita , in cujus locum nova non eft
affumpta, planum conftituere reperitur, fi nempe incognitarum neutra in fe ipfam in æquatione ducta fit;
fin fecus, dimidio tantùm ejus quantitatis , quæ planum conftituit cum incognita, in cujus locum affump-
tio facta eft, cafu utroque juxta differentem affectio-
nem per figna $+$ vel $-$, quæ præfiguntur iifdem illis
quantitatibus, ita ordinatis, ut cum incognitis ab eadem æquationis parte reperiantur. Quo facto, & reï-
terato, ubi opùs, fi ad formulas Parabolarum, capite
fecundo expofitas perventum non fuerit , ad aliquem
quatuor fuprapofitorum cafuum reducta erit æquatio,
ac proinde ipfi convenientem Locum determinare ac
defcribere, per ea quæ fuperiùs explicata funt, haud difficile erit.

Exemplum reductionis æquationum ad formulam Theorematis XI.

Si æquatio fuerit $yx - cx + hy \infty ee$: affumpto $z \infty y - c$, &
$v \infty x + h$, erit $z + c \infty y$, & $v - h \infty x$.

Vnde fi fecundùm Regulam ubique in æquatione loco y fub-
ftituatur $z + c$, erit $zx + cx - cx + hz + hc \infty ee$, five $zx + hz$
$+ hc \infty ee$; ac rurfus fi loco ipfius x fubrogetur $v - h$, erit $zv -$
$hz + hz + hc \infty ee$, id eft, $zv \infty ee - hc$. aut, (fi loco termini
$ee - hc$, qui in totum cognitus eft, fcribatur ff) $zv \infty ff$. Et
apparet æquationem reductam effe ad formulam Theorema-
tis XI, ac proinde Locum quæfitum effe Hyperbolam.

Ad cujus fpecificam determinationem ac defcriptionem efto
in appofita figura initium ipfius x immutabile punctum A, atque
eadem x per rectam A E indefinitè fe extendere intelligatur, fitque angulus, quem y & x comprehendunt, æqualis angulo E A K

Pars II.   N n   aut

[282]

or its supplement. Because $z = y - c$, and if we suppose that y rises above AE, we have next also to draw the straight line KB above AE and parallel to it, so that the part of the straight line AK and of all its parallels intercepted between AE and KB, such as AK are equal to the known c. Furthermore, because $v = x + h$, BK has to be produced through K up to G so that $KG = h$. In this way G will be the center of the curve and GB one of its asymptotes [3.9] and the other one will be parallel to AK, like GH. Henceforth, if, according to the Rule included in the aforementioned Theorem XI, we measure GC (equal to the known f) on the straight line GB and if we draw CF equal to GC and parallel to the straight line AK or GH, and if we describe the hyperbola FD through F with GB and GH as its asymptotes or with GB as an asymptote and GF as its axis, then, I say, the curve FD will be the required locus.

Indeed, if we select a point at random on it, such as D, and if we draw DE parallel to AK, which intersects the straight line KB at B and if this DE is called y, then DB or $DE - EB = y - c = z$. However, also GB or $AE + GK = x + h = v$. As, by the nature of a hyperbola, the rectangle GBD equals the square on GC, therefore also $zv = f^2$.

If we substitute $y - c$ for z, $x + h$ for v, and $e^2 - ch$ for f^2, then

$$yx - cx + hy - ch = e^2 - ch, \text{ i.e. } yx - cx + hy = e^2.$$

Which was proposed.

282 ELEM. CVRVARVM

aut ejufdem ad duos rectos fupplemento. Deinde, quoniam z eft
$\infty y - c$, fi y fupra lineam A E exfurgere concipiatur, ducenda
quoque eft fupra eandem recta K B ipfi A E parallela; ita ut pars
rectæ A K, omniumque ipfi æquidiftantium, inter A E & K B intercepta, veluti A K, æquetur c cognitæ. Porrò, quoniam v eft
$\infty x + h$, producenda eft ipfa B K per K ufque ad G, ita ut K G fit

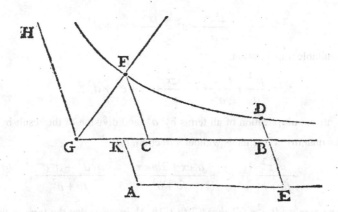

∞h. Quo facto, erit G centrum ipfius curvæ, & G B una Afym-
ptotωn, eritque altera ipfi A K parallela, ut G H. Vnde fi juxta
Regulam prædicti Theorematis X I in recta G B fumatur G C æ-
qualis f cognitæ, ducaturque C F eidem G C æqualis, ac parallela rectæ A K vel G H, atque per punctum F, Afymptotis G B &
G H, five Afymptoto G B atque ad axem G F, Hyperbola defcribatur F D: dico curvam F D fore Locum quæfitum.

Sumpto enim in ea puncto utcunque, veluti D, ductâque D E
ipfi A K parallelâ, quæ fecet rectam K B in B, fi eadem D E vocetur y, erit D B five D E — E B $\infty y - c$, id eft, z. Eft autem &
G B five A E + G K $\infty x + h$, hoc eft, v. Quare cum ex natura
Hyperboles rectangulum G B D æquetur G C quadrato, erit
quoque $z\,v \infty ff$. aut reftitutis $y - c$ loco ipfius z, & $x + h$ in locum ipfius v, atque $ee - ch$ loco ff, erit $yx - cx + hy - ch \infty$
$ee - ch$, hoc eft, $yx - cx + hy \infty ee$. Quod erat propofitum.

Exem-

[283]

Examples of the reduction of equations to the form of Theorems XII and XIII

If the equation is $y^2 + 2(bxy/a) + 2cy = (fx^2/a) + ex + d^2$ and if we put $z = y + (bx/a) + c$, then $y = z - (bx/a) - c$; if we substitute this for y and its square for y^2 [3.11] and if we delete all terms that cancel each other, then

$$z^2 - \frac{b^2 x^2}{a^2} - \frac{2bcx}{a} - c^2 = \frac{fx^2}{a} + ex + d^2 .$$

And after a suitable transposition,

$$z^2 = \frac{fx^2}{a} + \frac{b^2 x^2}{a^2} + \frac{2bcx}{a} + ex + c^2 + d^2 .$$

This means after multiplication of all terms by a^2 and division of the result by $fa + b^2$ in order that the quantity x^2 remains without a fraction, that

$$\frac{a^2 z^2}{fa + b^2} = x^2 + \frac{a^2 ex + 2abcx}{fa + b^2} + \frac{a^2 d^2 + a^2 c^2}{fa + b^2} .$$

If we put next $v = x + [(a^2 e + 2abc)/(2fa + 2b^2)]$ in order that the term in the equation in which x is found in the first degree also totally vanishes, we will have $x = v - [(a^2 e + 2abc)/(2fa + 2b^2)]$. If we substitute this for x and its square for x^2, and if we delete those terms that cancel each other, then the equation will be reduced to the required form [3.12].

But let us write $2h$ instead of $(a^2 e + 2abc)/(fa + b^2)$, in order to avoid a too long operation so that the equation becomes

$$\frac{a^2 z^2}{fa + b^2} = x^2 + 2hx + \frac{a^2 d^2 + a^2 c^2}{fa + b^2} .$$

If we then put $v = x + h$ or $x = v - h$ and if we substitute this for x in the equation and its square for x^2, then we will have

$$\frac{a^2 z^2}{fa + b^2} = v^2 - h^2 + \frac{a^2 d^2 + a^2 c^2}{fa + b^2} .$$

Lɪʙ. II. Cᴀᴘ. III.   283

Exempla reductionis æquationum ad formulam Theorematis XII & XIII.

Si æquatio sit $yy + \frac{2bxy}{a} + 2cy \infty \frac{fxx}{a} + ex + dd$, assumpto $z \infty y + \frac{bx}{a} + c$, erit $y \infty z - \frac{bx}{a} - c$, eoque substituto in locum ipsius y, atque ejusdem quadrato loco yy, sublatisque iis, quæ se invicem destruunt, erit $zz - \frac{bbxx}{aa} - \frac{2bcx}{a} - cc \infty \frac{fxx}{a} + ex + dd$.

Et factâ congruâ transpositione, $zz \infty \frac{fxx}{a} + \frac{bbxx}{aa} + ex + \frac{2bcx}{a} + dd + cc$, hoc est, multiplicatis omnibus æquationis terminis per aa, productoque diviso per $fa + bb$; ut quantitas xx absque fractione remaneat, fiet $\frac{aazz}{fa+bb} \infty xx + \frac{aaex + 2abcx}{fa+bb} + \frac{aadd + aacc}{fa+bb}$.

Deinde assumpto $v \infty x + \frac{aae + 2abc}{2fa + 2bb}$, ut terminus quoque æquationis, in quo x unius dimensionis reperitur, planè evanescat, habebitur $x \infty v \frac{-aac - 2abc}{2fa + 2bb}$. Quo substituto in locum ipsius x, atque ejusdem quadrato loco xx, ablatisque iis quæ se invicem tollunt, reducta erit æquatio ad formulam requisitam. At verò ut vitetur prolixior operatio loco $\frac{aae + 2abc}{fa + bb}$ scribatur $2h$, ita ut fiat æquatio $\frac{aazz}{fa+bb} \infty xx + 2hx + \frac{aadd + aacc}{fa+bb}$. Tum assumpto $v \infty x + h$ seu $x \infty v - h$, eoque substituto loco x in æquatione, ac ejusdem quadrato loco xx : erit $\frac{aazz}{fa+bb} \infty vv - hh + \frac{aadd + aacc}{fa+bb}$. Vnde apparet, ante omnia hîc esse considerandum, utrum hh sit majus quàm $\frac{aadd + aacc}{fa+bb}$, an contra. si enim majus sit, erit casus Theorematis XII; sin contra, erit casus Theorematis XIII. Ponatur itaque primò majus, ac proinde æquatio formulæ Theorematis XII. Et constat exinde Locum quæsitum Hyperbolam esse.

Ad cujus peculiarem determinationem esto in apposita figura ipsius x initium immutabile A punctum, eademque x in linea A E. ab A versùs E indefinitè se extendere intelligatur; sitque angulus

Nn 2   quem

[283, cont.]

From this it is clear that we have to examine in the first place whether h^2 is greater than $(a^2d^2 + a^2c^2)/(fa + b^2)$ or the opposite inequality holds [3.13].

Indeed, if h^2 is greater, then it will be an example of Theorem XII, but otherwise it will be an example of Theorem XIII. Let us first suppose that h^2 is greater and, hence, that the equation has the form of Theorem XII [3.14]. Then it is certain that the required locus is a hyperbola.

Let, for its accurate determination, in the added figure point A be the immutable initial point of x and let us suppose that this x extends indefinitely along the line AE from A toward E and let the angle

[284]

enclosed by x and y be equal to the angle EAK or to its supplement. Because $z = y + c + (bx/a)$ and if we suppose that y rises above the line AE, just as ED, we have first to draw the straight line KL below AE and parallel to AE, so that the part of the straight line AK and of all its parallels intercepted between the aforementioned AE and KL, just as AK, EL, etc., are equal to the known c. Next we first produce LK up to H so that $KH = h$, whence HL, supposed to be indefinite, equals $x + h$, that is v. Then we draw the segment HG through point H parallel to AK so that KH is to HG as a is to b. After these preparations the following holds: if we draw the straight line GKB through the points G an K, then the parts of all parallels to AK intercepted between KL and KB (e.g. just as LB) will have the same ratio to the parts of KL intercepted between the same parallels and point K (e.g. just as LK) as the ratio of b to a. This means that KL is to LB as a is to b, because both proportions are as KH is to HG. As KL or AE, supposed to be indefinite, equals x, then LB and any parallel segment intercepted between KL and KB will therefore be equal to bx/a and so every line rising above AE, that can act as the unknown y and that has been produced up to the straight line KB, as DB or $DE + EL + LB$, is equal to $y + c + (bx/a)$, that is z. Hence it is clear, according to the Rule, that we have to choose GB as a diameter of the hyperbola. The ordinate-wise applied lines have to be parallel to AK or DB

284   E L E M. C V R V A R V M

quem x & y comprehendunt æqualis angulo E A K aut ipſius ad duos rectos ſupplemento. Porrò quoniam $z \infty y + c + \frac{bx}{a}$, ſi y ſupra lineam A E exſurgere intelligatur, ut E D, ducenda primùm eſt infra eandem A E recta K L ipſi A E parallela ; ita ut pars rectæ A K omniumque ipſi æquidiſtantium inter prædictas A E & K L intercepta, veluti A K, E L, &c. æquetur c cognitæ. Deinde productâ L K uſque ad H, ita ut K H ſit ∞ h, ideoque H L indefinitè ſumpta ∞ $x + h$, hoc eſt, v, ducatur per H punctum recta H G ipſi A K parallela, ita ut K H ad H G ſit, ut a ad b. Quo facto , ſi per puncta G & K recta agatur linea G K B, habebunt omnium ipſi A K parallelarum partes, quæ inter K L & K B in-

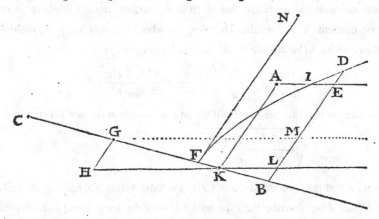

tercipiuntur (ut, exempli gratiâ, L B), ad partes ipſius K L, inter eaſdem parallelas & punctum K interceptas (ut, verbi gratiâ, L K) eandem rationem, quæ eſt inter b & a: hoc eſt, erit ut a ad b, ita K L ad L B, cum utraque ſit ut K H ad H G. Ideoque cum K L ſive A E indefinitè ſumpta ſit ∞ x, erit L B, ut & quælibet ipſi parallela, inter K L & K B intercepta, ∞ $\frac{bx}{a}$: ac proinde omnis linea ſupra A E rectam exſurgens, quæ poſſit eſſe y incognita, ad rectam K B producta , veluti D B ſive D E + E L + L B erit ∞ $y + c + \frac{bx}{a}$, hoc eſt, z. Vnde apparet, juxta Regulam lineam G B ſumendam eſſe pro Hyperbolæ diametro, ad quam ordinatim applicatæ ſint ipſi A K ſeu D B parallelæ : eritque ejuſdem

[285]

and point G will be the center of this hyperbola. Because, on the basis of the foregoing, the triangle KHG is wholly known (the sides KH and HG and the angle at H enclosed by them being known), therefore the ratio of KH to KG, that is the ratio of GM (supposed to pass through G and to be parallel to KL) to GB will also be known. Let it be as a is to i. Because GM or HL, supposed to be indefinite, is equal to v, therefore GB, supposed to be indefinite as well. That is any part of the diameter intercepted between the center and the ordinate-wise applied lines, will be iv/a. The square of this intercept, however, constitutes one term of the equation according to the formula of the Rule. Therefore the equation has to be reduced by multiplication or by division or by both in this way that in this equation the square i^2v^2/a^2 is also found [3.15].

In order that this be done in a strict rule [3.15], the aforementioned square of the segment GB, supposed to be indefinite (that is i^2v^2/a^2), has to be divided by the term of the equation in which v^2 is found, either on its own or preceded by another fraction, and the whole equation has to be multiplied by the quotient that has been found. For example: if in the previous example (p. [283], line 10 from the bottom, transl.) i^2v^2/a^2 is divided by v^2, the quotient i^2/a^2 results. Therefore the whole equation has to be multiplied by i^2 and the result has to be divided by a^2, so that we get

$$\frac{i^2z^2}{fa+b^2} = \frac{i^2v^2}{a^2} - \frac{i^2h^2}{a^2} + \frac{i^2d^2+i^2c^2}{fa+b^2}.$$

If, according to the Rule, we make half the latus transversum GF or GC equal to

$$\sqrt{\frac{i^2h^2}{a^2} - \frac{i^2d^2+i^2c^2}{fa+b^2}}$$

and the ratio of the latus transversum CF to the latus rectum FN equal to the ratio of i^2 to $fa+b^2$ and if we describe the hyperbola FD with the same latera and with the already found diameter and center, intersecting the straight line AE or KA (being produced) at I, then, I say, the curve ID is the required locus.

Indeed, if we select a point at random on this curve, say D and if we draw DE parallel to AK and if we produce DE so that it intersects the straight line KL at L and meets the diameter GB at B and if DE is called y, then $DB = z$, on the basis of what we stated before. However, as we already remarked $GB = iv/a$ and by hypothesis

L I B. II. C A P. III.   285

dem Hyperbolæ centrum G punctum. At verò cum ex ante dictis triangulum K H G omnino sit cognitum, utpote lateribus K H & H G anguloque ad H sub iisdem comprehenso notis, erit quoque cognita ratio lateris K H ad K G, hoc est, ipsius G M (quæ per G ipsi K L æquidistans intelligitur) ad G B, quæ sit ut a ad i. Quare cum G M seu H L indefinitè sit v, G B quoque indefinitè concepta, hoc est, quælibet diametri portio, inter centrum & ordinatim applicatas intercepta, erit $\frac{iv}{a}$. Cujus quidem interceptæ quadratum cum juxta formulam Regulæ unum æquationis terminum constituat, per multiplicationem aut divisionem, vel per utramque ita reducatur æquatio, ut in eadem quoque idem quadratum, nimirum $\frac{iivv}{aa}$ inveniatur. Quod quidem ut certâ methodo fiat, prædictum quadratum rectæ G B indefinitè conceptæ, hoc est, $\frac{iivv}{aa}$, dividatur per æquationis terminum, in quo vv sive simpliciter, sive aliâ fractione affectum invenitur, ac per inventum quotientem tota æquatio multiplicetur. ut in supra posito exemplo, si $\frac{iivv}{aa}$ dividatur per vv, fiet quotiens $\frac{ii}{aa}$. quare tota æquatio multiplicanda est per ii, productumque dividendum per aa, ita ut fiat $\frac{iizz}{fa+bb} \,\infty\, \frac{iivv}{aa} - \frac{iibb}{aa} + \frac{iidd+iicc}{fa+bb}$. Vnde si juxta Regulam semi-latus transversum fiat G F vel G C ∞ $\sqrt{\frac{iibb}{aa} - \frac{iidd-iicc}{fa+bb}}$, atque ratio transversi lateris C F ad rectum F N, ut ii ad $fa+bb$, & iisdem lateribus, diametroque ac centro jam inventis Hyperbole describatur F D, secans rectam A E vel K A productam in I : dico curvam I D esse Locum quæsitum.

Sumpto enim in eadem curva puncto utcunque, veluti D, ductâque D E ipsi A K parallelâ, eâque productâ ut secet rectam K L in L, & diametro G B occurrat in B, si eadem D E vocetur y, erit ex ante dictis D B ∞ z. Est autem, ut jam annotatum, G B ∞ $\frac{iv}{a}$, atque ex hypothesi G F seu G C ∞ $\sqrt{\frac{iibb}{aa} - \frac{iidd-iicc}{fa+bb}}$, ideoque B C ∞ $\frac{iv}{a} + \sqrt{\frac{iibb}{aa} - \frac{iidd-iicc}{fa+bb}}$, ac B F ∞ $\frac{iv}{a} -$

Nn 3 $\sqrt{\frac{iibb}{aa}}$

[285, cont.]

$$GF \text{ or } GC = \sqrt{\frac{i^2 h^2}{a^2} - \frac{i^2 d^2 + i^2 c^2}{fa + b^2}}$$

and so

$$BC = \frac{iv}{a} + \sqrt{\frac{i^2 h^2}{a^2} - \frac{i^2 d^2 + i^2 c^2}{fa + b^2}},$$

and

[286]

$$BF = \frac{iv}{a} - \sqrt{\frac{i^2 h^2}{a^2} - \frac{i^2 d^2 + i^2 c^2}{fa + b^2}},$$

and the rectangle CBF satisfies

$$CBF = \frac{i^2 v^2}{a^2} - \frac{i^2 h^2}{a^2} + \frac{i^2 d^2 + i^2 c^2}{fa + b^2}.$$

Because by the nature of a hyperbola, NF is to FC or $fa + b^2$ is to i^2, as the square on DB (that is z^2) is to the aforementioned rectangle CBF [3.16], therefore

$$\frac{i^2 z^2}{fa + b^2} = \frac{i^2 v^2}{a^2} - \frac{i^2 h^2}{a^2} + \frac{i^2 d^2 + i^2 c^2}{fa + b^2}.$$

If we now on both sides multiply by a^2 and divide by i^2, then we will get

$$\frac{a^2 z^2}{fa + b^2} = v^2 - h^2 + \frac{a^2 d^2 + a^2 c^2}{fa + b^2},$$

If we next substitute $x + h$ for v, then

$$\frac{a^2 z^2}{fa + b^2} = x^2 - 2hx + \frac{a^2 d^2 + a^2 c^2}{fa + b^2}.$$

In the same way substitution of $(ea^2 + 2bca)/(fa + b^2)$ for $2h$ gives

$$\frac{a^2 z^2}{fa + b^2} = x^2 - \frac{ea^2 x + 2bcax}{fa + b^2} + \frac{a^2 d^2 + a^2 c^2}{fa + b^2}.$$

If we furthermore multiply the whole by $fa + b^2$ and if we divide by a^2, we will have

$$z^2 = \frac{fx^2}{a} + \frac{b^2 x^2}{a^2} + ex + \frac{2bcx}{a} + d^2 + c^2.$$

Finally, if we substitute $y + x/a + c$ for z, delete all terms that cancel each other and if we arrange all terms as usual, then we will get

286   E L E M.   C V R V A R V M

$\sqrt{\dfrac{iibb}{aa} \dfrac{-iidd-iicc}{fa+bb}}$, & rectangulum C B F ∞ $\dfrac{iivv}{aa}$ − $\dfrac{iibb}{aa}$ +

$\dfrac{iidd+iicc}{fa+bb}$. Hinc cum ex natura Hyperboles N F ad F C, seu

$fa+bb$ ad ii sit, ut D B quadratum, hoc est, zz, ad prædictum

rectangulum C B F: erit $\dfrac{iizz}{fa+bb}$ ∞ $\dfrac{iivv}{aa}$ − $\dfrac{iibb}{aa}$ + $\dfrac{iidd+iicc}{fa+bb}$,

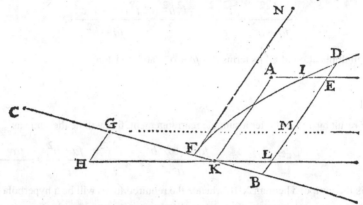

Multiplicetur jam utrinque per aa, & dividatur per ii, eritque

$\dfrac{aazz}{fa+bb}$ ∞ $vv-hh$ + $\dfrac{aadd+aacc}{fa+bb}$. Dein restituto $x+h$ loco v,

exurget $\dfrac{aazz}{fa+bb}$ ∞ $xx+2hx$ + $\dfrac{aadd+aacc}{fa+bb}$; itemque $\dfrac{eaa+2bca}{fa+bb}$

loco $2h$, exurget $\dfrac{aazz}{fa+bb}$ ∞ xx + $\dfrac{eaax+2bcax}{fa+bb}$ + $\dfrac{aadd+aacc}{fa+bb}$.

Porrò multiplicatis omnibus per $fa+bb$ iisque divisis per aa,

habebitur zz ∞ $\dfrac{fxx}{a}$ + $\dfrac{bbxx}{aa}$ + ex + $\dfrac{2bcx}{a}$ + $dd+cc$. Ac deni-

que restituto $y+\dfrac{bx}{a}+c$ loco ipsius z, expunctisque quæ se in-

vicem destruunt ac omnibus ritè ordinatis, fiet yy + $\dfrac{2bxy}{a}$ + $2cy$

∞ $\dfrac{fxx}{a}$ + ex + dd. Quod erat propositum.

At verò ponatur secundò hh minus quàm $\dfrac{ddaa+ccaa}{fa+bb}$, & su-

pra posita æquatio $\dfrac{iizz}{fa+bb}$ ∞ $\dfrac{iivv}{aa}$ − $\dfrac{iibb}{aa}$ + $\dfrac{ddii+ccii}{fa+bb}$, quæ,

multiplicatis omnibus ejusdem terminis per $fa+bb$, ac producto

diviso

[286, cont.]

$$y^2 + \frac{2bxy}{a} - 2cy = \frac{fx^2}{a} + ex + d^2.$$

Which was proposed.

Let us secondly [3.17] suppose that h^2 is smaller than $\dfrac{d^2a^2 + c^2a^2}{fa + b^2}$ and let the above proposed equation be

$$\frac{i^2z^2}{fa + b^2} = \frac{i^2v^2}{a^2} - \frac{i^2h^2}{a^2} + \frac{i^2d^2 + i^2c^2}{fa + b^2}.$$

After multiplication of all its terms by $fa + b^2$, and division

[287]

of the result by i^2 and after a suitable rearrangement, this will be the same as

$$z^2 - d^2 - c^2 + \frac{fah^2 + b^2h^2}{a^2} = \frac{i^2v^2}{a^2} \cdot \frac{fa + b^2}{i^2} = \frac{fav^2 + b^2v^2}{a^2}.$$

This is the form of Theorem XIII, whence the required locus will be a hyperbola [3.18].

For its accurate determination and description we first draw (as in the preceding figure) the lines AE, AK, KL, KH, HG and GKB. Then, as before, G will be the center, but the diameter will not at all be on line GK but, according to the Rule [3.19], on line HG produced toward G. The lines ordinate-wise applied to this diameter have to be parallel to GKB and, according to the same Rule, half the transverse diameter, namely GF or GC will be equal to

$$\sqrt{d^2 + c^2 - \frac{fah^2 + b^2h^2}{a^2}}$$

and the ratio of the diameter to the parameter will be as $fa + b^2$ is to i^2. If we put CF to FN as $fa + b^2$ to i^2 (FN being parallel to GKB), then FN will be the parameter. Hence, if the hyperbola FD is described with G as its center, CF as its transverse diameter and FN as its parameter (the hyperbola which intersects this AE or the product KA in I), then the curve ID will be the required locus

Indeed, if we select a point at random on this curve, e.g D, and if we draw DB parallel to AK (or GF) and DM parallel to GB

LIB. II. CAP. III.   287

divifo per *i i*, factâque decenti tranfpofitione, eadem cum fequenti

$$z\,z - d\,d - c\,c + \frac{fabh + bbhh}{aa} \,\infty\, \frac{iivv}{aa}\, \text{multip. per} fa + bb \text{ ac di-}$$

vif. per *i i*, id eft, $\infty \dfrac{favv + bbvv}{aa}$. erit formulæ Theorema-

tis X I I I, unde Locus quæfitus iterum erit Hyperbola. Ad cujus
fpecificam determinationem & defcriptionem, poftquam ut in
præcedenti figura ductæ funt lineæ A E, A K, K L, K H, H G, &
G K B: erit quidem, ut fupra, G centrum, at verò non erit diameter in linea G K, fed , juxta Regulam, in linea H G producta

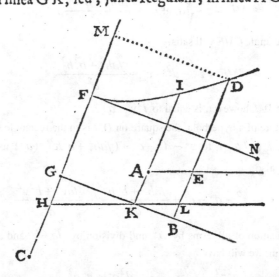

ad partes G, ad quam ordinatim applicatæ fint ipfi G K B parallelæ , eritque juxta eandem Regulam dimidium tranfverfæ diametri, nempe G F vel G C, æquale $\sqrt{dd + cc\dfrac{-fabh - bbhh}{aa}}$, ac
ratio diametri ad parametrum ut $fa + bb$ ad *i i*. Quare fi fiat, ut
$fa + bb$ ad *i i*, ita C F ad F N, quæ quidem F N ipfi G K B æquidiftans fit , erit F N parameter : ac proinde fi centro G tranfversâ
diametro C F & parametro F N Hyperbola defcribatur F D, fecans ipfam A E vel K A productam in I, erit I D curva Locus
quæfitus.

Sumpto enim in eadem curva puncto utcunque, veluti D, ductâque D B ipfi A K (five G F), & D M ipfi G B parallelâ, fi

E D

[288]

and if *ED* is called y, then, as before, *DB* or $MG = z$ and *BG* or $\breve{D}M = DM = iv/a$. Because *GF* or *GC* is equal to

$$\sqrt{d^2 + c^2 - \frac{fah^2 + b^2h^2}{a^2}} \; ,$$

therefore

$$CM = z + \sqrt{d^2 + c^2 - \frac{fah^2 + b^2h^2}{a^2}}$$

and

$$MF = z - \sqrt{d^2 + c^2 - \frac{fah^2 + b^2h^2}{a^2}} \; .$$

And so the rectangle *CMF* will satisfy

$$CMF = z^2 - d^2 - c^2 + \frac{fah^2 + b^2h^2}{a^2} \; .$$

The square on *DM*, however, is equal to i^2v^2/a^2 .

By the nature of a hyperbola, the square on *DM* is to the rectangle *CMF* as *FN* is to *FC* [3.20], that is, i^2v^2/a^2 is to $z^2 - d^2 - c^2 + (fah^2 + b^2h^2)/(a^2)$ as i^2 is to $fa + b^2$.

Therefore also

$$z^2 - d^2 - c^2 + \frac{fah^2 + b^2h^2}{a^2} = \frac{fav^2 + b^2v^2}{a^2} \; .$$

After multiplication of all terms by a^2 and division by $fa + b^2$ and after transposition of the known term, we will have

$$\frac{a^2z^2}{fa + b^2} = v^2 - h^2 + \frac{d^2a^2 + c^2a^2}{fa + b^2} \; .$$

If we next substitute $x + h$ for v, $(ea^2 + 2bca)/(fa + b^2)$ for $2h$, and $y + (bx/a + c)$ for z, and if we delete all terms that cancel each other and rearrange all terms as usual, then we will have

$$y^2 + \frac{2bxy}{a} + 2cy = \frac{fx^2}{a} + ex + d^2 \; .$$

Which was to be determined and to be demonstrated.

288   E L E M.   C V R V A R V M

E D vocetur y, erit, ut supra, D B five M G ∞z, & B G five D M $\infty \frac{iv}{a}$. Cumque fit G F vel G C $\infty \sqrt{dd+cc\dfrac{-fabh-bbhh}{aa}}$,

erit C M $\infty z + \sqrt{dd+cc\dfrac{-fabh-bbhh}{aa}}$, & M F $\infty z -$

$\sqrt{dd+cc\dfrac{-fabh-bbhh}{aa}}$, ac propterea rectangulum C M F ∞

$zz-dd-cc\dfrac{+fabh+bbhh}{aa}$. Est autem D M quadratum $\infty \dfrac{iivv}{aa}$.

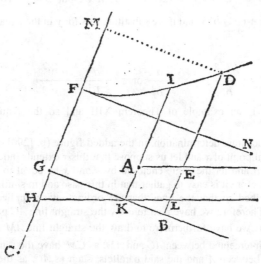

Quare cum ex natura Hyperboles fit ut F N ad F C, ita D M quadratum ad C M F rectangulum, hoc eft, ut ii ad $fa+bb$, ita $\dfrac{iivv}{aa}$ ad $zz-dd-cc\dfrac{+fabh+bbhh}{aa}$: erit quoque $z z-dd-cc$ $\dfrac{+fabh+bbhh}{aa} \infty \dfrac{favv+bbvv}{aa}$. Et multiplicatis omnibus per aa, ac divifis per $fa+bb$, factâque tranfpofitione cogniti termini, erit $\dfrac{aazz}{fa+bb} \infty vv-hh\dfrac{+ddaa+ccaa}{fa+bb}$. Dein reftitutis $x+h$ loco v, $\dfrac{eaa+2bca}{fa+bb}$ loco $2h$, atque $y+\dfrac{bx}{a}+c$ loco ipfius z, expunctisque quæ fe invicem deftruunt ac omnibus ritè ordinatis, fiet $yy+\dfrac{2bxy}{a}+2cy \infty \dfrac{fxx}{a}+ex+dd$. Quod determinandum, demonftrandumque erat.

Si

[289]

Suppose that the equation is

$$x^2 + 2ay = \frac{2byx}{a} \quad \text{or} \quad x^2 - \frac{2byx}{a} + 2ay = 0.$$

If, according to the Rule, we put $v = x - by/a$, then $x = v + by/a$ and if we substitute this for x and its square for x^2 and if we remove the terms that cancel each other, then $v^2 - (b^2 y^2 / a^2) + 2ay = 0$ and after a suitable transposition $v^2 = (b^2 y^2 / a^2) - 2ay$, this is, after multiplication of all terms of the equation by a^2 and after division of the result by b^2, then $(a^2 v^2 / b^2) = y^2 - (2a^3 y / b^2)$. If we put $z = y - (a^3 / b^2)$, then we will next have $y = z + (a^3 / b^2)$ and if we substitute this for y in the equation and its square for y^2, then

$$\frac{a^2 v^2}{b^2} = z^2 - \frac{a^6}{b^4} \quad \text{or} \quad z^2 - \frac{a^6}{b^4} = \frac{a^2 v^2}{b^2}.$$

This, however, is an example of Theorem XIII and so the required locus will be a hyperbola.

Let, for its accurate determination, in the added figure (p. [290], transl.) point A be the immutable initial point of x and let us suppose that this x extends indefinitely along line AB from A toward B and let the angle enclosed by y and x be equal to the angle ABE. From what we stated before it is easy to gather that in this case and in similar cases the hyperbola has to be described so that the lines that are ordinate-wise applied to its diameter, are parallel to AB. Therefore we have next to draw the straight line AC parallel to BE. Because $v = x - (by/a)$, we have furthermore to draw the straight line AM so that the parts of all parallels to AB intercepted between AC and AM, as CM, have the same ratio to the parts of AC intercepted between A and the said parallels, such as AC, as the ratio of a to b. This means that AC is to CM as a is to b. Hence, if AC or BE (supposed to be indefinite) is called y, then CM and similar segments will be equal to by/a and the diameter of the hyperbola to be described will be situated on the said AM. Furthermore the following holds because $z = y - (a^3 / b^2)$: if we subtract AF $(= a^3 / b^2)$ from AC, then FC (supposed to be indefinite) equals z and if we draw FN parallel to AB, then N will be the center. Because the ratio of ND (drawn parallel to FC and equal to it) to the segment DM (and the ratio of other similar segments) is known, namely as a to b and because the angle NDM

L I B. II. C A P. III. · 289

Si æquatio fit $xx + 2 ay \infty \frac{2bxy}{a}$,aut $xx - \frac{2byx}{a} + 2 ay \infty 0$.

Affumpto juxta Regulam $v \infty x - \frac{by}{a}$, erit $x \infty v + \frac{by}{a}$, eoque fubftituto in locum ipfius x, ejufdemque quadrato loco xx, fublatif-
que iis quæ fe invicem deftruunt, erit $vv - \frac{bbyy}{aa} + 2 ay \infty 0$. &,
factâ congruâ tranfpofitione, $vv \infty \frac{bbyy}{aa} - 2 ay$; hoc eft, multiplicatis omnibus æquationis terminis per aa, productoque divifo
per bb, $\frac{aavv}{bb} \infty yy - \frac{2a^3 y}{bb}$. Dein, affumpto $z \infty y - \frac{a^3}{bb}$, habebitur $y \infty z + \frac{a^3}{bb}$, eoque fubftituto in æquatione loco ipfius y, atque ipfius quadrato loco yy, erit $\frac{aavv}{bb} \infty zz - \frac{a^6}{b^4}$, five $zz - \frac{a^6}{b^4}$
$\infty \frac{aavv}{bb}$. Qui quidem cafus eft Theorematis 13^{tii}, ac proinde
Locus quæfitus erit Hyperbola.

Ad cujus itaque peculiarem determinationem efto in appofita
figura ipfius x initium immutabile A punctum, eademque x in linea A B ab A versùs B indefinitè fefe extendere intelligatur, fitque angulus, quem x & y comprehendunt, æqualis angulo A B E.
Deinde, quoniam ex antedictis facilè colligitur Hyperbolam hoc
cafu & fimilibus ita effe defcribendam, ut ordinatim ad ejus diametrum applicatæ fint ipfi A B æquidiftantes, ductâ rectâ A C ipfi
B E parallelâ, quoniam $v \infty x - \frac{by}{a}$, ducenda porrò eft recta
A M; ita ut omnium ipfi A B parallelarum partes, inter A C &
A M interceptæ, veluti C M, ad partes ipfius A C inter A & dictas parallelas interceptas, veluti A C, eandem rationem habeant, quæ eft inter b & a; hoc eft, ut fit quemadmodum a ad b,
ita A C ad C M. Vnde fi A C feu B E indefinitè fumpta voce-
tur y, erit C M & fimiles $\infty \frac{by}{a}$, ac defcribendæ Hyperboles diameter in dicta A M. Porrò, quoniam $z \infty y - \frac{a^3}{bb}$, fi ab A C auferatur A F $\infty \frac{a^3}{bb}$: erit F C indefinitè fumpta ∞z, &, ductâ F N
ipfi A B parallelâ, N centrum. Ac proinde, cum ratio ductæ N D
ipfi F C æquidiftantis & æqualis ad rectam D M aliarumque fimilium fit cognita, nempe ut a ad b, fitque itidem notus angulus

Pars II.　　　　　　　O o　　　　　　　N D M,

[290]

enclosed by them is known as well (it is namely equal to the given or chosen angle *ABE*), therefore the ratio of *ND* to *NM* and the ratio of other similar segments is also known, let us say as the ratio of the known *a* to *e* which is also known. Because *ND* or *FC* (supposed to be indefinite) is expressed by *z*, therefore *NM* (supposed to be indefinite as well) will be equal to ez/a, and as, according to the formula of the Rule, its square has to constitute one term of the equation, the above equation (p. [289], line 9 and 10 from the top, transl.), therefore, has to be multiplied by e^2 and the result has to be divided by a^2 so that we get

$$\frac{e^2 z^2}{a^2} - \frac{e^2 a^4}{b^4} = \frac{e^2 v^2}{b^2}.$$

If, hereafter, according to the Rule, we make half the latus rectum *NG* or *NH* equal to ea^2/b^2 and the ratio of the latus transversum to the latus rectum as e^2 tot b^2 and if we describe the hyperbola *GE* with these latera and with the diameter and the center already found, then, I say, the curve *GE* will be the required locus.

Indeed, if we select a point at random on it, say *E*, and if we draw *EB* at the angle *ABE* equal to the given or chosen angle, and *EC* parallel to *AB*, intersecting the diameter *AM* at *M*, then the following holds. If *EB* (that is *AC*) is called *y*, then, as before, $CM = by/a$ and so *ME* (or $AB - CM$) $= x - (by/a) = v$. However, as we remarked before, $NM = ez/a$ and, by hypothesis, *NG* (or *NH*) $= ea^2/b^2$ and, therefore, $HM = (ez/a) + (ea^2/b^2)$ and $MG = (ez/a) - (ea^2/b^2)$ and so the rectangle *HMG* satisfies

290   E L E M.   C V R V A R V M

N D M, fub iifdem comprehenfus, utpote æqualis dato vel affumpto angulo A B E, erit quoque ratio N D ad N M aliarumque fimilium nota, quæ fit ut *a* cognitæ ad *e* itidem cognitam.

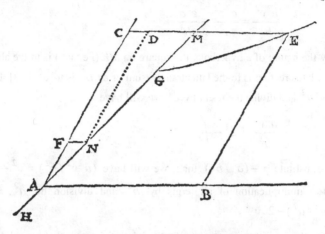

Hinc cum N D feu F C indefinitè fumpta exprimatur per *z*, erit
N M itidem indefinitè fumpta $\infty \frac{ez}{a}$, cujus quidem quadratum
cum juxta formulam Regulæ unum æquationis terminum conftituere debeat, multiplicanda eft fuprafcripta æquatio per *e e*,
productumque dividendum per *a a*, ita ut fiat $\frac{eezz}{aa} - \frac{eca^4}{b^4} \infty \frac{ecvv}{bb}$.
Quo peracto, fi juxta Regulam femi-latus tranfverfum fiat N G
vel N H $\infty \frac{eaa}{bb}$, ac ratio tranfverfi lateris ad rectum, ut *e e* ad *b b*;
iifdemque lateribus ac diametro & centro jam inventis Hyper-
bole defcribatur G E : dico curvam G E effe Locum quæfitum.
    Sumpto enim in ea puncto utcunque, veluti E, ductâque E B
in angulo A B E, dato vel affumpto æquali, nec non E C ipfi A B
parallelâ, fecante diametrum A M in M; fi eadem E B, hoc eft,
A C, vocetur *y*, erit, ut fupra, C M $\infty \frac{by}{a}$, ac proinde M E, five
A B—C M, $\infty x - \frac{by}{a}$, hoc eft, *v*. Eft autem, ut fuperiùs annotatum, N M $\infty \frac{ez}{a}$, atque ex hypothefi N G feu N H $\infty \frac{eaa}{bb}$, ideoque H M $\infty \frac{ez}{a} + \frac{eaa}{bb}$, & M G $\infty \frac{ez}{a} - \frac{eaa}{bb}$, ac proinde rectangulum

[291]

$$HMG = \frac{e^2 z^2}{a^2} - \frac{e^2 a^4}{b^4}.$$

From this it follows

$$\frac{e^2 v^2}{b^2} = \frac{e^2 z^2}{a^2} - \frac{e^2 a^4}{b^4}$$

because, by the nature of a hyperbola, the square on *ME* (i.e. v^2) is to the above rectangle *HMG* as the latus rectum is to the latus transversum or as b^2 is to e^2 [3.3]. If we multiply all terms by a^2 and divide the result by e^2 we will have

$$\frac{a^2 v^2}{b^2} = z^2 - \frac{a^6}{b^4}.$$

If we next substitute $y - (a^3 / b^2)$ for z, we will have $(a^2 v^2 / b^2) = y^2 - (2a^3 / b^2)y$, and so, after multiplication of all terms by b^2 and division by a^2, we will have $v^2 = (b^2 y^2 / a^2) - 2ay$.

If we finally substitute $x - by / a$ for v, delete the terms that cancel each other and arrange all terms as usual, the result will be $x^2 + 2ay = 2byx / a$. Which was proposed.

PROBLEM II

Proposition 16

Two points being given, it is required to find a third point with the property that the segments drawn from this point to each of the two given points differ by a given distance and to determine and to describe the locus to which the required point belongs.

Let the two given points be *A* and *B* and let it be required to find a third point, say *C*, so that the drawn segments *CA* and *CB* differ by a given distance *FG* or *AD* [3.21].

Because in this problem the angle has not been given, we suppose it to be a right angle for convenience's sake. Therefore, we suppose that from point *C* the perpendicular, say *CE*, has been dropped to the segment *AB* (which connects the given points) or, if necessary, to its produced part. Next, according to the Rule, we suppose the unknown and indeterminate *AE* and *EC*, enclosing an assumed angle *AEC*, as known and determinate [3.22]. Let the former, *AE* to be sure, be called *x* and let the latter, namely *EC*, be named *y*; let, however, *AB* itself or the known distance of the given points be called *a* and let the given *FG* or *AD* be expressed by *b*. Now either $BE = AE - AB$ (if point *B* falls between *A* and *E*) or $BE = AB - AE$ (if point *E* falls between *A* and *B*), so $BE = |x - a|$ [3.23] and

gulum H M G ∞ $\frac{eezz}{aa}$ — $\frac{eea^4}{b^4}$: hinc cum ex natura Hyperboles
fit ut latus rectum ad tranfverfum, five ut bb ad ee, ita M E quadratum, id eft, vv, ad prædictum rectangulum H M G : erit
$\frac{eevv}{bb}$ ∞ $\frac{eezz}{aa}$ — $\frac{eea^4}{b^4}$, &, multiplicatis omnibus terminis per aa,

factoque per ee divifo, $\frac{aavv}{bb}$ ∞ zz — $\frac{a^5}{b^4}$. Dein reftituto y — $\frac{a^3}{bb}$

in locum ipfius z, exurget $\frac{aavv}{bb}$ ∞ yy — $\frac{2a^3y}{bb}$; adeoque, multi-

plicatis omnibus per bb, factoque divifo per aa, habebitur
vv ∞ $\frac{bbyy}{aa}$ — $2ay$. Denique reftituto x — $\frac{by}{a}$ in locum ipfius v,
expunctisque iis quæ fe invicem deftruunt, atque omnibus ritè
ordinatis, fiet $xx + 2ay$ ∞ $\frac{2byx}{a}$. Quod fuit propofitum.

P ʀ ᴏ ʙ ʟ ᴇ ᴍ ᴀ II.

Propofitio 16.

Datis duobus punctis tertium invenire, à quo ad
bina data ductæ rectæ lineæ dato differant intervallo,
locumque determinare ac defcribere, quem quæfitum
punctum contingat.

Sint data duo puncta A & B, oporteatque invenire tertium, utputa C, ita nempe ut ductæ rectæ C A, C B differant dato intervallo F G feu A D.

Quoniam in quæftione angulus datus non eft, quò facilior fit
operatio, affumatur rectus, ideoque à puncto C in rectam A B,
quæ data puncta conjungit, productam, fi opùs fuerit, intelligatur demiffa perpendicularis, ut C E; tum, fuppofitis, juxta Regulam, A E & E C incognitis atque indeterminatis, affumptum
angulum A E C comprehendentibus, tanquam cognitis ac determinatis, earum prior, nimirum A E, vocetur x, ac pofterior,
nempe E C, nominetur y, ipfa autem A B, feu datorum punctorum cognita diftantia, vocetur a, & data F G five A D exprimatur per b. Hinc cum B E five (fi punctum B cadat inter A & E)
A E — A B, aut (fi punctum E inter A & B cadat) A B — A E fit

[292]

$AC = \sqrt{x^2 + y^2}$, but $BC = \sqrt{x^2 - 2ax + a^2 + y^2}$ and $AC - AD = BC$; therefore, the equation will be

$$\sqrt{x^2 + y^2} - b = \sqrt{x^2 - 2ax + a^2 + y^2} .$$

After a suitable operation in order that both members of the equation are freed from the radical sign and after transferring what has to be transferred, the equation will be

$$4b^2 y^2 = 4a^2 x^2 - 4b^2 x^2 - 4a^3 x + 4b^2 ax + a^4 - 2b^2 a^2 + b^4.$$

After division by $4a^2 - 4b^2$ [3.24] we will have

$$\frac{b^2 y^2}{a^2 - b^2} = x^2 - ax + \tfrac{1}{4}a^2 - \tfrac{1}{4}b^2.$$

If, according to the Rule, we next put $v = x - \tfrac{1}{2}a$, then $x = v + \ x = v + \tfrac{1}{2}a$ and if we substitute this value for x and its square for x^2, and if we delete the terms that cancel each other, then

$$\frac{b^2 y^2}{a^2 - b^2} = v^2 - \tfrac{1}{4}b^2$$

This, however, is an example of Theorem XII of this book and so the required locus will be a hyperbola.

Now we substituted v for $x - \tfrac{1}{2}a$; so the following holds: if we measure $AH = \tfrac{1}{2}a$ from A towards E, then, according to the Rule, H will be the center and half the transverse diameter will be $\sqrt{\tfrac{1}{4}b^2}$, that is $\tfrac{1}{2}b$ (namely HG on one side and HF on the other side). In this way the transverse diameter FG equals b and because CE has been applied as a perpendicular to the diameter HE, this transverse diameter is also the axis. However, the ratio of the transverse diameter to the parameter or rather the ratio of the square on the transverse diameter to the square on the second diameter will be as b^2 to $a^2 - b^2$. So it will not be difficult to describe the hyperbola itself by means of what has been explained in the second and in the last chapter of the first book [3.25].

$\infty\, x = a$, & $A\,C \infty \sqrt{xx+yy}$, at $B\,C \infty \sqrt{x\,x - 2\,a\,x + a\,a + yy}$; fitque $\overline{A\,C} - A\,D \infty \overline{B\,C}$: æquatio erit

$\sqrt{xx+yy} - b \infty \sqrt{xx - 2\,a\,x + a\,a + yy}$, factâque operatione convenienti, ut utraque æquationis pars à figno radicali liberetur, & tranfpofitis tranfponendis, erit

$4\,bb\,yy \infty 4\,aa\,xx - 4\,bb\,xx - 4\,a^3\,x + 4\,bb\,ax + a^4 - 2\,bbaa + b^4$.

Vnde factâ divifione per $4\,aa - 4\,bb$ habebitur $\frac{bbyy}{aa-bb} \infty xx - ax + \frac{1}{4}aa - \frac{1}{4}bb$. Deinde affumpto juxta Regulam $v \infty x - \frac{1}{2}a$, erit $x \infty v + \frac{1}{2}a$, ideoque fubftituto hoc valore in locum ipfius x, atque ejufdem quadrato loco $x\,x$, expunctifque iis quæ fe invi-

Fig. 1.

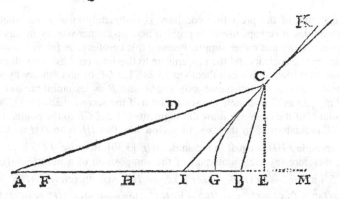

cem deftruunt, erit $\frac{bbyy}{aa-bb} \infty vv - \frac{1}{4}bb$. Qui quidem cafus eft Theorematis 12^{mi} hujus libri, ac proinde Locus quæfitus erit Hyperbola. Cumque v affumpta fit pro $x - \frac{1}{2}a$, fi ab A versùs E fumatur $A\,H \infty \frac{1}{2}a$, erit, juxta Regulam, H centrum, & femidiameter tranfverfa (puta H G ab una, & H F ab altera parte,) $\infty \sqrt{\frac{1}{4}bb}$, id eft, $\frac{1}{2}b$; ita ut diameter tranfverfa F G (quæ quidem, ob applicatam C E ad diametrum H E perpendicularem, tranfverfus quoque axis eft,) fit ∞b. Ratio autem tranfverfæ diametri ad parametrum, feu quadrati tranfverfæ ad quadratum fecundæ diametri, erit ut bb ad $aa - bb$. Vnde per ea quæ libri primi capitibus fecundo & ultimo expofita funt Hyperbolam ipfam defcribere haud difficile erit. Porrò cum quadratum femidiame-

[293]

Furthermore, as the square on the transverse semi-diameter is equal to $\frac{1}{4}b^2$, the square on half the second diameter will be $\frac{1}{4}a^2 - \frac{1}{4}b^2$. But as FB, or $BH + HF$, is equal to $\frac{1}{2}a + \frac{1}{2}b$ and BG, or $BH - HG$, equals $\frac{1}{2}a - \frac{1}{2}b$, the rectangle FBG will therefore be equal to $\frac{1}{4}a^2 - \frac{1}{4}b^2$, namely equal to the square on half the second diameter or, as the ancients said, equal to a quarter of "the figure" applied to the transverse axis [3.26]. Therefore, the points A and B are exactly the points which in general are called foci or umbilical points of opposite hyperbolas. Thus it appears that the right conclusions have been drawn from the assumptions. [3.27]

Corollary 1

If, from a point selected at random on a hyperbola, segments are drawn to both umbilici, then the greatest of them will exceed the smallest by the length of the transverse axis.

Although the truth of the preceding corollary is completely clear from what we stated before, yet we think it perhaps useful to prove it here in another way by means of a unique demonstration, shorter and rather simple, because this corollary, as far as I know, has been demonstrated by the ancients and the moderns up to this time only by many digressions and by a long concatenation of difficult theorems [3.28]. Let GC be an arbitrary hyperbola with H as its center, FG as its transverse axis and A and B as its umbilici, and so that the rectangle FBG just as GAF equals the square on half the second diameter [3.29]. From an arbitrary point C of the curve we draw the segments CA and CB to the points A and B, and we apply CE ordinate-wise to the axis, in such a way that HE is to HM as HF is to HA. Thus[1] the rectangle FHM is equal to rectangle AHE [3.30]. Because[2] FEG is to CE^2 as HF^2 is to GAF, therefore (by the counterpart of the composition of a ratio [3.30]), FEG is to $(FEG + CE^2)$ as HF^2 is to $(HF^2 + GAF)$, i.e.[3] to HA^2 and, therefore,[4] $(HF^2 + FEG)$, or[5] HE^2, is to $(HA^2 + FEG + CE^2)$ as HF^2 is to HA^2. However, also[6] HE^2 is to HM^2 as HF^2 is to HA^2. Therefore,[7] $HM^2 = HA^2 + FEG + CE^2$; which means after adding HF^2 or HG^2 on both sides

---

[1] VI, 16.

[2] By hypothesis and by Lib. I, Prop. 10, p. [196].

[3] II, 6.

[4] Because the sum of all antecedents is to the sum of all consequents as an antecedent is to its consequent and by V, 12.

[5] II, 6.

[6] By construction and by VI, 22.

[7] V, 9 and V, 11.

Lɪʙ. II. Cᴀᴘ. III.   293

diametri tranſverſæ ſit ꝏ ¼ *b b*, erit quadratum ſemi-ſecundæ diametri ꝏ ¼ *a a* — ¼ *b b*. Atqui cum F B ſive B H + H F ſit ꝏ ½ *a* + ½ *b*,
& B G ſive B H — H G ꝏ ½ *a* — ½ *b*, erit quoque rectangulum
F B G ꝏ ¼ *a a* — ¼ *b b*, nempe ꝏ quadrato ſemi-ſecundæ diametri,
ſive, ut Veteres loquebantur, æquale quadranti figuræ ad tranſverſum axem factæ : ideoque puncta A & B ea ipſa ſunt, quæ vul-
go oppoſitarum Hyperbolarum Foci ſive Vmbilici nuncupantur.
Vnde apparet, ex præmiſſis rectè inferri, quæ ſequuntur.

Corollarium ɪ.

Si ab aſſumpto utcunque in Hyperbola puncto ad u-
trumque Vmbilicum rectæ ducantur, earum major minorem longitudine tranſverſi axis ſuperabit.

Etiamſi veritas præcedentis Corollarii ex antedictis omnino
conſtet, cum tamen illud à Veteribus, Recentioribuſvè, quòd
ſciam, non niſi per multas ambages longâque difficilium Theorematum concatenatione hactenus demonſtratum ſit : id ipſum
hîc demonſtratione unicâ, & quidem breviore ſatiſque ſimplici,
aliter oſtendiſſe non inutile fortè judicabitur.

Eſto igitur Hyperbola quælibet G C, cujus centrum H , tranſ-
verſus axis F G, atque Vmbilici A & B, adeoque rectangulum
F B G ut & G A F ſemi-ſecundæ diametri quadrato æquale. Ductis autem ab aſſumpto quolibet curvæ puncto C ad puncta A &
B rectis C A, C B, ordinatim ad axem applicetur C E, ſiatque ut
H F ad H A, ita H E ad H M, ideoque [1] A H E rectangulo æ-
quale rectangulum F H M. Vnde cum ſit [2], ut

H F *q* ad G A F, ita F E G ad C E *q* : erit quoque, per compoſ. rationis contrariam, ut

H F *q* ad (H F *q* + G A F, id eſt [3], ad) H A *q*; ita F E G ad F E G
+ C E *q*; adeoque [4] ut

H F *q* ad H A *q*, ita (H F *q* + F E G ſive [5]) H E *q* ad H A *q* +
F E G + C E *q*. Eſt autem quoque [6], ut

H F *q* ad H A *q*, ita H E *q* ad H M *q*. Quocirca [7].

H M *q* ꝏ H A *q* + F E G + C E *q*; hoc eſt, addito utrinque H F *q*
ſeu H G *q*, erit

Oo 3   HM

Margin notes:
[1] *per 16 ſexti.*
[2] *ex hy-poth. & per 10 primi hujus.*
[3] *per 6 ſecundi.*
[4] *cum ſit ut una antecedentium ad unam conſeq., ita omnes antecedentes ad omnes*

conseq. per 12 quinti. [5] per 6 ſecundi. [6] ex conſtructione & per 22 ſexti. [7] per 9 & 11 quinti.

[294]

$$HM^2 + HF^2 = HA^2 + (HF^2 + FEG) \text{ [i.e.}^1 \ HE^2] + CE^2$$

or

$$HM^2 + HG^2 = HB^2 + (HF^2 + FEG) \text{ [i.e.}^1 \ HE^2] + CE^2.$$

If we add or subtract equal terms on both sides of the equation, namely 2*FHM*, respectively 2*GHM*, on one side and 2*AHE*, respectively 2*BHE*, on the other side, then[2]

$$FM^2 = (AE^2 + CE^2) \text{ [i.e.}^3 \ AC^2] \text{ and likewise}^4$$

$$GM^2 = (BE^2 + CE^2) \text{ [i.e.}^5 \ BC^2].$$

As from this follows *FM = AC* and *GM = BC* and as *FG* is the difference between *FM* and *GM*, it is also clear, therefore, that the greatest of *AC* and *BC* exceeds the smallest by the length of this *FG*, namely the transverse axis. Which was to be demonstrated.

Corollary 2

If one draws straight lines from an arbitrary point of a hyperbola to either umbilicus, then the line that bisects the angle enclosed by these straight lines, touches the curve at this point and conversely.

Indeed, if the straight line *ICK* that bisects the angle *ACB*, does not touch the hyperbola at point *C*, then it will, if possible, intersect the hyperbola and so at least one of its points, say *K*, will lie within the hyperbola.

---

[1] II, 6. [2] II, 4. [3] I, 47. [4] II, 7. [5] I, 47.

294   E L E M.   C V R V A R V M

1 *per 6 secundi.*

$$HMq + \begin{cases} HFq \\ \text{seu} \\ HGq \end{cases} \infty \begin{cases} HAq \\ \text{seu} \\ HBq \end{cases} + (HFq + FEG, \text{i.e.}^1)HEq, + CEq.$$

Hinc additis vel sublatis ab utraque æquationis parte æqualibus,

2 *per 4 secundi.*

$$\text{nimirum} \begin{cases} FHM \\ \text{seu} \quad \text{bis ab una, } \& \\ GHM \end{cases} \begin{cases} AHE \\ \text{seu} \quad \text{bis ab altera parte:} \\ BHE \quad \text{erit}^2 \end{cases}$$

3 *per 47 primi.*
4 *per 7 secundi.*
5 *per 47 primi.*

$FMq \infty (AEq + CEq, \text{id est }^3) ACq;$ itemque ^{4}
$GMq \infty (BEq + CEq, \text{id est }^5) BCq.$ Cumque propterea
FM sit $\infty AC; \& GM \infty BC;$ sitque ipsarum $FM \& GM$ dif-
ferentia FG, manifestum est ipsarum quoque $AC \& BC$ majorem superare minorem, ejusdem FG, nempe axis transversi, longitudine. Quod demonstrandum erat.

Corollarium 2.

Ductis à quolibet Hyperbolæ puncto ad utrumque
Vmbilicum rectis , quæ angulum iis comprehensum
bifariam dividit linea curvam in eodem puncto contingit; & conversim.

Si enim quæ angulum A C B bifariam dividit recta I C K non
contingat Hyperbolam in C puncto, secet eandem, si fieri potest,

Fig. 1.

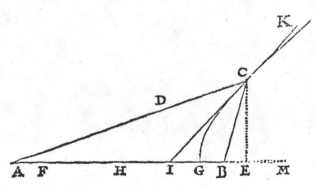

atque ita saltem aliquo sui puncto, veluti K, intra Hyperbolam sit.
Tum

[295]

If we then draw KB, KD [3.31] and KA (the last of them intersecting the hyperbola at L from which BL has been drawn to B), the following holds: because in the triangles DCK and BCK the sides DC and CK are respectively equal to BC and CK, and as they enclose equal angles[1] [3.32], therefore the base DK will also be equal to the base BK. As moreover, according to the preceding corollary, AL exceeds BL by the distance AD and as this also holds for AK and the segments BL and LK (if taken together) and as BK (and so also KD) is smaller than BL and LK (if taken together), therefore AK will exceed KD by a distance greater than AD, which means that AK will be greater than the two segments KD and DA (if taken together) [3.33]. As this is very absurd[2], the straight line ICK does not intersect the hyperbola, but it touches this hyperbola at point C.

---

[1] Indeed, because by hypothesis the angles ACI and BCI are supposed to be equal, therefore their adjacent angles ACK and BCK will also be equal by I, 13.

[2] I, 10.

L I B. II. C A P. II.   295

Tum ductis K B, K D, & K A (quarum posterior Hyperbolam
secet in L, à quo ad B ducta sit B L), cum in triangulis D C K,
B C K latera D C, C K lateribus B C, C K utrumque utrique,

· *Fig.* I I.

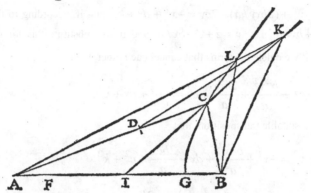

circa [1] æquales angulos, æqualia sint, erit quoque basis D K basi B K æqualis. Cumque porrò, juxta Corollarium præcedens,
A L ipsam L B, ideoque & A K rectas B L, L K, simul sumptas,
superet intervallo A D; sitque B K, ideoque & K D, ipsis B L,
L K simul sumptis minor : per consequens A K eandem K D ma-
jori longitudine quàm est A D excedet, id est, ipsa A K binis re-

[margin right column:]
[1] Cum e-
nim ex
hypothe-
si anguli
A C I &
B C I æ-
quales
ponan-
tur, erunt
quoque
anguli
A C K &
B C K,
qui ipsis
sunt
deinceps,
per 13
primi æ-
quales.

Fig. I I I.

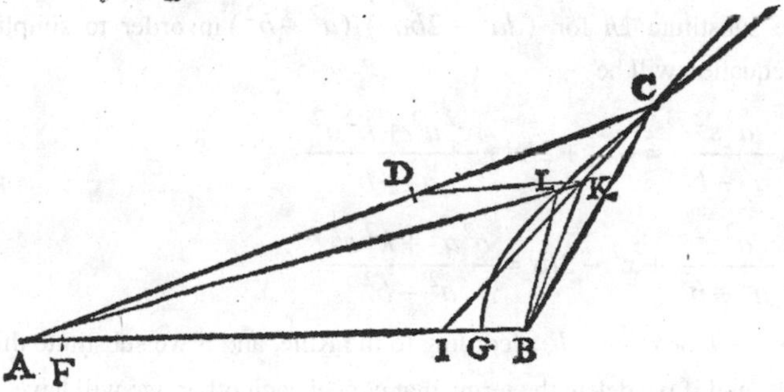

ctis K D, D A simul sumptis major erit. Quod cum absurdissi-
mum sit [2], non secat Hyperbolam recta I C K, sed eandem contingit in C puncto. Cumque non possit in eodem puncto C alia

[2] *per* 20
primi.

recta.

[296]

As no other straight line than ICK[1] can touch the hyperbola at this point C, it is conversely clear that a straight line that touches the hyperbola at C, also bisects the angle ACB (the text erroneously gives ABC, transl.).

Example of the reduction of equations to the form of Theorem XIV

If the equation is $y^2 + (2bxy/a) - 2cy = -x^2 + dx + k^2$ and if, according to the Rule, we put $z = y - c + bx/a$, i.e. $y = z + c - bx/a$ and if we substitute this for y and its square for y^2 and if we delete the terms that cancel each other, then

$$z^2 - \frac{b^2 x^2}{a^2} + \frac{2bcx}{a} - c^2 = -x^2 + dx + k^2,$$

which means, after a suitable transposition, that

$$z^2 = -x^2 + \frac{b^2 x^2}{a^2} + dx - \frac{2bcx}{a} + c^2 + k^2$$

or
$$z^2 = \frac{-a^2 x^2 + b^2 x^2}{a^2} + \frac{dax - 2bcx}{a} + c^2 + k^2.$$

If, however, we suppose a to be greater than b and if we multiply all terms of the equation by a^2 and if we divide the result by $a^2 - b^2$ [3.34], in order that the quantity x^2 is found without a fraction, then

$$\frac{a^2 z^2}{a^2 - b^2} = -x^2 + \frac{da^2 x - 2bacx}{a^2 - b^2} + \frac{c^2 a^2 + k^2 a^2}{a^2 - b^2}.$$

If, furthermore, we substitute $2h$ for $(da^2 - 2bac)/(a^2 - b^2)$ in order to simplify the operation, then the equation will be

$$\frac{a^2 z^2}{a^2 - b^2} = -x^2 + 2hx + \frac{c^2 a^2 + k^2 a^2}{a^2 - b^2}$$

or
$$\frac{a^2 z^2}{a^2 - b^2} + x^2 - 2hx = \frac{c^2 a^2 + k^2 a^2}{a^2 - b^2}.$$

Hence, if we put $v = x - h$ or $x = v + h$, according to the Rule, and if we substitute this for x and its square for x^2 and if we delete the terms that cancel each other, we will have

$$\frac{a^2 z^2}{a^2 - b^2} + v^2 - h^2 = \frac{c^2 a^2 + k^2 a^2}{a^2 - b^2}.$$

This means after a suitable transposition

---

[1] Lib.I, Prop. 6, Cor. 3, p [191].

recta Hyperbolam contingere quàm I C K ¹, manifestum est conversim, eam, quæ Hyperbolam in C contingit, angulum quoque
A B C bifariam dividere.

Exemplum reductionis æquationum ad formulam Theorematis XIV.

Si æquatio sit $yy + \frac{2bxy}{a} - 2cy \infty - xx + dx + kk$, assum-

pto juxta Regulam $z \infty y - c + \frac{bx}{a}$, hoc est, $y \infty z + c - \frac{bx}{a}$,

eoque substituto in locum ipsius y, ejusdemque quadrato loco yy,

sublatisque iis, quæ se invicem destruunt, erit $zz - \frac{bbxx}{aa} + \frac{2bcx}{a}$

$- cc \infty - xx + dx + kk$. id est, factâ decenti transpositione,

erit $zz \infty - xx + \frac{bbxx}{aa} + dx - \frac{2bcx}{a} + cc + kk$ sive

$zz \infty \frac{-aaxx+bbxx}{aa} + \frac{dax-2bcx}{a} + cc + kk$. Supposito au-

tem a majore quàm b, ac multiplicatis omnibus æquationis terminis per aa, productoque diviso per $aa - bb$, ut quantitas xx

absque fractione inveniatur, erit $\frac{aazz}{aa-bb} \infty - xx \frac{+daax-2bacx}{aa-bb}$

$+ \frac{ccaa+kkaa}{aa-bb}$. Iam verò si facilioris operationis gratiâ loco

$\frac{daa-2bac}{aa-bb}$ substituatur $2h$: erit æquatio $\frac{aazz}{aa-bb} \infty - xx + 2hx$

$+ \frac{ccaa+kkaa}{aa-bb}$, aut $\frac{aazz}{aa-bb} + xx - 2hx \infty \frac{ccaa+kkaa}{aa-bb}$.

Hinc si juxta Regulam assumatur $v \infty x - h$ sive $x \infty v + h$, atque
hoc in locum ipsius x, ejusque quadratum loco xx substituatur,
ac expungantur quæ se invicem destruunt, habebitur $\frac{aazz}{aa-bb}$

$+ vv - hh \infty \frac{ccaa+kkaa}{aa-bb}$. Hoc est, factâ decenti transpositio-

ne, erit $\frac{aazz}{aa-bb} \infty - vv + hh + \frac{ccaa+kkaa}{aa-bb}$. Atque ita appa-

ret æquationem esse reductam ad formulam Theorematis XIV,
ideoque Locum quæsitum aut Ellipsin aut Circuli circumferentiam existere. Rursus verò facilioris operationis ergo loco

$\frac{aa}{aa-bb}$ scribatur $\frac{l}{g}$, & loco $hh + \frac{ccaa+kkaa}{aa-bb}$ scribatur ff, ita

ut æquatio sit talis $\frac{lzz}{g} \infty ff - vv$.

Ad

[296, cont.]

$$\frac{a^2 z^2}{a^2 - b^2} = -v^2 + h^2 + \frac{c^2 a^2 + k^2 a^2}{a^2 - b^2}.$$

In this way it is clear that the equation has been reduced to the form of Theorem XIV and thus, that the required locus appears to be either an ellipse or the circumference of a circle.

Let us again, to simplify the operation, write l/g instead of $a^2/(a^2 - b^2)$ and f^2 instead of $h^2 + (c^2 a^2 + k^2 a^2)/(a^2 - b^2)$ so that the equation is as follows: $lz^2/g = f^2 - v^2$.

[297]

Let, however, for an accurate determination and description of the aforementioned locus, in the next figure point A be the immutable initial point of x and let us suppose that this x extends indefinitely along the line AE from A toward E and let the given or chosen angle enclosed by y and x, be equal to the angle EAK or to its supplement. Now $z = y - c + (bx/a)$, so if we suppose that y rises above the line AE we have also to draw the straight line KL above AE and parallel to it, so that the part of the straight line AK and the parts of all its parallels intercepted between the aforesaid AE and KL, such as AK, EL, etc., are equal to the known c.

Next we have to draw the straight line KB through point K, below the straight line KL at such an angle that the parts of all parallels to AK intercepted between KL and KB (such as LB) have the same ratio to the parts of KL intercepted between these parallels and point K (such as LK) as the ratio between b and a, which means that KL is to LB as a is to b [3.35]. Hence, if we put KL or AE (assumed to be indefinite) equal to x, then LB and all its parallels intercepted between KL and KB will be equal to bx/a. Hence it is clear from what we stated before, that the diameter will be situated on the straight line KB,

Lib. II. Cap. III.   297

Ad peculiarem autem prædicti Loci determinationem ac defcriptionem efto in appofita figura ipfius *x* initium immutabile A punctum, atque eadem *x* fe in linea A E ab A versùs E indefinitè extendere intelligatur, fitque angulus datus vel affumptus, quem *y* & *x* comprehendunt, æqualis angulo E A K vel ejufdem ad duos rectos complemento. Hinc quoniam $z \infty y - c + \frac{bx}{a}$, fi *y* fupra lineam AE exfurgere intelligatur, ducenda quoque eft fupra ipfam recta K L eidem parallela, ita ut pars rectæ A K omniumque ipfi æquidiftantium inter prædictas A E & K L intercepta, veluti A K,

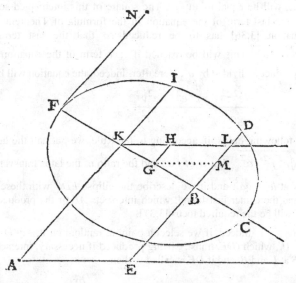

E L, &c. æquetur *c* cognitæ: ac deinde per punctum K infra rectam K L ducenda eft recta K B in tali angulo, ut rectarum omnium ipfi A K parallelarum partes, quæ inter K L & K B intercipiuntur (veluti L B) ad partes ipfius K L, inter eafdem parallelas & punctum K interceptas (ut verbi gratiâ L K) eandem habeant rationem, quæ eft inter *b* & *a*, hoc eft, ut fit uti *a* ad *b*, ita K L ad L B. Atque ita pofitâ K L five A E, indefinitè fumptâ, ∞x, L B omnesque ipfi parallelæ inter K L & K B interceptæ erunt $\frac{bx}{a}$. Vnde ex prædictis conftat diametrum fore in recta K B,

Pars II.   P p   ad

[298]

whose ordinate-wise applied lines are parallel to AK. As $v = x - h$ we have furthermore to subtract KH from the segment KL or AE so that $KH = h$ and, therefore, HL (likewise assumed to be indefinite) is equal to $x - h$ or v. Next we have to draw HG through point H, parallel to AK and intersecting the just found diameter at G; then this point of intersection G will be the center of the required ellipse. Because in the similar triangles KHG and KLB the ratio of the side KH to HG or the ratio of KL to LB is known as well as the angle enclosed by these sides, which is namely equal to the given or chosen angle EAK, the ratio of the side KH to the side KG or the ratio of KL to KB is therefore also known. Let us put this ratio equal to the ratio of the known a to an e which is known as well. Therefore, the following holds. Because HL or GM (understood to be parallel to HL), assumed to be indeterminate, is equal to v, therefore GB, likewise assumed to be indeterminate, that is an arbitrary part of the diameter intercepted between the center and an arbitrary ordinate-wise applied line, will be equal to ev / a. The square of this intercepted segment (GB, transl.) constitutes the last term of the equation in the formula of Theorem XIV; therefore, the above equation [3.36] has to be reduced so that the last term of this equation becomes e^2v^2 / a^2. This will be reached if each term of the equation is multiplied by e^2 and if the product is divided by a^2. Hereafter, indeed, the equation will be

$$\frac{le^2z^2}{ga^2} = \frac{f^2e^2}{a^2} - \frac{e^2v^2}{a^2}.$$

Hence the following holds: if, according to the Rule, we put half the latus transversum GF or GC at $\sqrt{e^2f^2 / a^2}$, that is ef / a, and the ratio of the latus transversum CF to the latus rectum FN at $le^2 : ga^2$, and if we describe the ellipse FDC with these latera and with the diameter and the center just found, which intersects AE or the produced AK in I, then the curve IDC will be the required locus [3.37].

Indeed, if we select a point at random on it, say D, and if we draw DE parallel to AK, which (DE, transl.), being produced if necessary, intersects the straight lines KL and KB at L and B, and if DE is called y, then

$$DB = DE - EL + LB = y - c + \frac{bx}{a} = z.$$

However, as we just now remarked, $GB = ev / a$ and by construction, GF or GC is equal to ef / a and so $FB = (ef / a) + (ev / a)$ and $BC = (ef / a) - (ev / a)$ and the rectangle FBC satisfies

$$FBC = \frac{f^2e^2}{a^2} - \frac{e^2v^2}{a^2}.$$

As by the nature of an ellipse [3.5]

298   ELEM. CVRVARVM

ad quam ordinatim applicatæ fint ipfi A K æquidiftantes. Iam verò
cum v fit ∞ $x-h$, à recta K L five A E auferenda eft K H, ita ut
eadem K H fit ∞ h, ideoque H L indefinitè quoque fumpta ∞
$x-h$ feu v. Deinde per punctum H ducenda eft H G ipfi A K
parallela, fecans inventam diametrum in G, eritque idem inter-
fectionis punctum G quæfitæ Ellipfeos centrum. Porrò quoniam
fimilium triangulorum K H G & K L B nota eft ratio lateris K H
ad H G five K L ad L B, ut & angulus fub iifdem lateribus con-
tentus, utpote æqualis angulo dato vel affumpto E A K, erit quoque nota ratio lateris K H ad latus K G five K L ad K B, quæ ponatur ut a cognitæ ad e itidem cognitam. Ideoque cum H L five
G M, quæ ipfi H L parallela intelligitur, indeterminatè fumpta
fit ∞ v, erit G B, fimiliter indeterminatè fumpta, hoc eft, quælibet diametri portio inter centrum & quamlibet ordinatim applicatam intercepta, $\infty \frac{ev}{a}$. Cujus quidem interceptæ quadratum
cum in formula Theorematis XIV ultimum æquationis termi-
num conftituat, æquatio fupra expofito modo ita reducatur, ut
terminus ejus extremus fiat $\frac{ecvv}{aa}$, id quod factum erit, fi finguli
æquationis termini multiplicentur per ee, productumque dividatur per aa. inde enim fequenti modo fe habebit æquatio
$\frac{leezz}{gaa} \infty \frac{ffee}{aa} - \frac{eevv}{aa}$. Hinc fi juxta Regulam femi-latus tranfverfum G F vel G C fiat $\infty \sqrt{\frac{ecff}{aa}}$, id eft, $\frac{ef}{a}$, & ratio tranfverfi lateris C F ad rectum latus F N, ut lee ad gaa, iifdemque lateri-
bus, ac diametro, centroque, modò inventis, Ellipfis defcribatur F D C, fecans rectam A E vel A K productam in I : erit curva
I D C Locus quæfitus.

Sumpto enim in ea puncto utcunque, veluti D, ductâque D E
ipfi A K parallelâ, ac fi opùs fit productâ ut fecet rectas K L &
K B in L & B, fi eadem D E vocetur y, erit D B, hoc eft, D E
$-$E L + L B $\infty y - c + \frac{bx}{a}$ feu z. Eft autem ut jam annotatum
eft G B $\infty \frac{ev}{a}$, atque ex conftructione G F vel G C $\infty \frac{ef}{a}$, ideoque
F B $\infty \frac{ef}{a} + \frac{ev}{a}$, & B C $\infty \frac{ef}{a} - \frac{ev}{a}$, ac rectangulum F B C $\infty \frac{ffee}{aa}$
$- \frac{eevv}{aa}$. Hinc cum ex natura Ellipfis fit ut N F ad F C, hoc eft,
ut

[299]

the square on *DB*, that is z^2, is to the aforementioned rectangle *FBC* as *NF* is to *FC*, that is as ga^2 is to le^2, therefore

$$\frac{le^2z^2}{ga^2} = \frac{f^2e^2}{a^2} - \frac{e^2v^2}{a^2},$$

which means, after multiplication of all terms by a^2 and after division by e^2,

$$\frac{lz^2}{g} = f^2 - v^2$$

and so, after substitution of $x - h$ for v, and

$$h^2 + \frac{c^2a^2 + k^2a^2}{a^2 - b^2}$$

for f^2, as well as $a^2/(a^2 - b^2)$ for l/g, that

$$\frac{a^2z^2}{a^2 - b^2} = h^2 + \frac{c^2a^2 + k^2a^2}{a^2 - b^2} - x^2 + 2hx - h^2,$$

that is

$$\frac{a^2z^2}{a^2 - b^2} + x^2 - 2hx = \frac{c^2a^2 + k^2a^2}{a^2 - b^2}.$$

Furthermore, if we substitute $(da^2 - 2bca)/(a^2 - b^2)$ for $2h$, then

$$\frac{a^2z^2}{a^2 - b^2} + x^2 + \frac{-da^2x + 2bcax}{a^2 - b^2} = \frac{c^2a^2 + k^2a^2}{a^2 - b^2},$$

which means, after multiplication by $a^2 - b^2$ and after division by a^2, that

$$z^2 + x^2 - \frac{b^2x^2}{a^2} - dx + \frac{2bcx}{a} = c^2 + k^2.$$

If, finally, we substitute $y - c + (bx/a)$ for z and if we delete the terms that cancel each other and if we arrange all terms as usual, then we will get

$$y^2 + \frac{2bxy}{a} - 2cy = -x^2 + dx + k^2.$$

Which was to be determined and to be demonstrated.

If the angle *AKB* would be a right angle so that the ordinate-wise applied lines, e.g. *DB*, *KI*, etc., are perpendicular to the diameter *KB*, and if at the same time *FN* would be equal to *FC*, then the aforementioned curve is a circle, as is clear from elementary considerations [3.38].

. ut gaa ad lee, ita D B quadratum, hoc eſt, zz ad prædictum re-
ctangulum F B C; erit $\frac{leezz}{gaa} \infty \frac{ffee}{aa} - \frac{eevv}{aa}$, id eſt, multiplica-

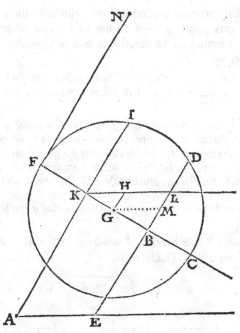

tis omnibus per aa,
ac diviſis per ee, erit
$\frac{lzz}{g} \infty ff - vv$, ideoque reſtituto $x - h$
loco v, atque $hh +$
$\frac{ccaa + kkaa}{aa - bb}$ loco ff,
ut & $\frac{aa}{aa - bb}$ loco $\frac{l}{g}$,
erit $\frac{aazz}{aa - bb} \infty hh +$
$\frac{ccaa + kkaa}{aa - bb} - xx +$
$2hx - hh$, hoc eſt,
$\frac{aazz}{aa - bb} + xx - 2hx$
$\infty \frac{ccaa + kkaa}{aa - bb}$. Porrò reſtituto
$\frac{daa - 2bca}{aa - bb}$ loco $2h$,
fiet $\frac{aazz}{aa - bb} + xx$
$\frac{-daax + 2bcax}{aa - bb} \infty$

$\frac{ccaa + kkaa}{aa - bb}$, id eſt, factâ multiplicatione per $aa - bb$ ac diviſione per aa, erit $zz + xx - \frac{bbxx}{aa} - dx + \frac{2bcx}{a} \infty cc + kk$.

Ac denique loco z factâ reſtitutione ipſius $y - c + \frac{bx}{a}$, deletiſque
iis quæ ſe invicem tollunt, ac omnibus ritè ordinatis, obtinebitur $yy + \frac{2bxy}{a} - 2cy \infty - xx + dx + kk$. Quod determinandum ac demonſtrandum erat.

Notandum porrò hîc eſt, quòd ſi angulus A K B foret rectus,
ac proinde ordinatim applicatæ, ut D B, K I, &c. ad diametrum
K B perpendiculares, ac ſimul F N æqualis F C, prædictam curvam fore Circulum, quemadmodum ex elementis perſpicuum eſt.

[300]

PROBLEM III

Proposition 17

Two points being given, it is required to find a third point with the property that the segments drawn from this point to each of the two given points are, taken together, equal to a given length, and to determine and to describe the locus to which the required point belongs.

Suppose that the points A and B are given [3.39]. Let it be required to find a third point, namely C, so that the drawn segments CA and CB, taken together, are equal to a given segment D.

Because in this problem the angle has not been given, let us suppose it to be a right angle, for convenience's sake. Therefore, we suppose that from point C the perpendicular, say CE, has been dropped to the segment AB, which connects the given points, if necessary to its produced part. Next, according to the Rule, we suppose the unknown and indeterminate AE and EC, enclosing the assumed angle AEC, as known and determinate. Let the former, AE to be sure, be called x and the latter, namely EC, be named y; let, however, AB itself or the known distance of the given points be called a and let the given D be expressed by b.

Now $BE = AB - AE$ (if point E falls between A and B) and $BE = AE - AB$ (if point B falls between A and E), so $BE = |a - x|$ ([3.23]) and

$$AC = \sqrt{x^2 + y^2} \text{ and } CB = \sqrt{a^2 - 2ax + x^2 + y^2}$$

and as $D - AC = CB$, the equation will be

$$b - \sqrt{a^2 - 2ax + x^2 + y^2} = \sqrt{a^2 - 2ax + x^2 + y^2}.$$

300   ELEM. CVRVARVM

PROBLEMA III.

Propositio 17.

Datis duobus punctis tertium invenire, à quo ad binum data ductæ rectæ lineæ simul sumptæ datæ longitudini æquales sint; locumque determinare ac describere, quem quæsitum punctum contingat.

Sint data duo puncta A & B, oporteatque invenire tertium, utputa C; ita nempe, ut ductæ rectæ CA, CB simul sumptæ æquales sint datæ rectæ lineæ D.

Quoniam in quæstione angulus datus non est, quò facilior fit operatio, assumatur rectus; ideoque à puncto C in rectam AB, quæ data puncta conjungit, productam, si opus fuerit,

Fig. 1.

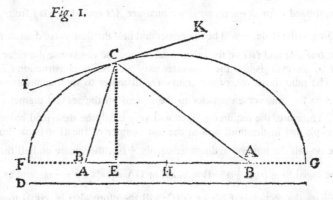

intelligatur demissa perpendicularis, ut CE. Tum suppositis, juxta Regulam, AE & EC incognitis atque indeterminatis assumptum angulum rectum AEC comprehendentibus tanquam cognitis ac determinatis, earum prior, nimirum AE, vocetur x, ac posterior, nempe EC, nominetur y; ipsa autem AB seu datorum punctorum distantia cognita appelletur a, & data D exprimatur per b. Hinc cum BE sive (si punctum E cadat inter A & B) AB—AE, aut (si punctum B inter A & E cadat) AE—AB sit ∞ $a = x$; atque AC ∞ $\sqrt{xx+yy}$; & CB ∞ $\sqrt{aa-2ax+xx+yy}$; sitque D—AC ∞ CB: æquatio erit $b-\sqrt{xx+yy}$ ∞ $\sqrt{aa-2ax+xx+yy}$; factâque operatione

[301]

After a suitable operation so that both members of the equation are freed of the radical sign and after transferring what has to be transferred, the equation will be

$$4b^2x^2 - 4a^2x^2 - 4b^2ax + 4a^3x = b^4 - 2b^2a^2 + a^4 - 4b^2y^2,$$

which means, after division by $4b^2 - 4a^2$ [3.40], that

$$x^2 - ax = \tfrac{1}{4}b^2 - \tfrac{1}{4}a^2 - \frac{b^2y^2}{b^2 - a^2}.$$

If, according to the Rule, we next put $v = x - (a/2)$, then $x = v + (a/2)$ and if we substitute this for x and its square for x^2 and if we delete the terms that cancel each other, then

$$v^2 = \tfrac{1}{4}b^2 - \frac{b^2y^2}{b^2 - a^2} \quad \text{or} \quad \frac{b^2y^2}{b^2 - a^2} = \tfrac{1}{4}b^2 - v^2 \qquad [3.41]$$

This, however, is an example of Theorem XIV and so the required locus is an ellipse. Now we substituted v for $x - \tfrac{1}{2}a$, so if we measure AH (equal to $\tfrac{1}{2}a$) from A toward E then, according to the Rule, H will be the center and half the transverse diameter equals $\tfrac{1}{2}b$ (e.g. HF on one side and HG on the other side) so that the transverse diameter FG is equal to b. This FG, however, is also the transverse axis because by construction CE is perpendicular to it. The ratio, however, of the transverse diameter to the parameter or the ratio of the square on the transverse diameter to the square on the second diameter will be as b^2 to $b^2 - a^2$. Therefore, the required ellipse will very easily be described by means of what has been explained in the third and in the last chapter of the first book. Furthermore as the square on half the transverse diameter equals $\tfrac{1}{4}b^2$, the square on half the second diameter will be equal to $\tfrac{1}{4}b^2 - \tfrac{1}{4}a^2$. But as FB or GA equals $\tfrac{1}{2}b + \tfrac{1}{2}a$ and as BG or AF equals $\tfrac{1}{2}b - \tfrac{1}{2}a$, the rectangle FBG or GAF, will therefore also be equal to $\tfrac{1}{4}b^2 - \tfrac{1}{4}a^2$; this means equal to the square on half the second diameter or, as the ancients said, equal to a quarter of the "figure" being applied to the transverse axis [3.42]. Therefore, the points A and B are precisely those that are generally called the foci or the umbilici of an ellipse. From this it is clear that the following is correctly inferred.

Corollary 1

The segments drawn from an arbitrary point on an ellipse to each of the umbilici are, taken together, equal to the transverse axis.

In the same way as it was shown above for the hyperbola that the difference of the segments CA and CB is equal to the transverse axis, it will be shown here that their sum is equal to the transverse axis. The proof, however, is not executed through addition and union,

ne decenti, ut utraque æquationis pars à signo radicali liberetur, & transpositis transponendis, erit

$$4\,bb\,xx - 4\,aa\,xx - 4bb\,ax + 4a^3x \infty b^4 - 2\,bb\,aa + a^4 - 4bbyy,$$

hoc est, factâ divisione per $4\,bb - 4\,aa$, erit

$xx - ax \infty \frac{1}{4}bb - \frac{1}{4}aa - \frac{bbyy}{bb - aa}$. Assumpto deinde juxta Regulam $v \infty x - \frac{1}{2}a$, erit $x \infty v + \frac{1}{2}a$, eâque substitutâ in locum ipsius x, ejusdemque quadrato loco xx, expunctisque iis quæ se invicem destruunt: erit $x\,x \infty \frac{1}{4}bb - \frac{bbyy}{bb - aa}$, sive $\frac{bbyy}{bb - aa} \infty \frac{1}{4}bb$ $- xx$. Qui quidem casus est Theorematis 13^{tii}, ac proinde Locus quæsitus Ellipsis. Cumque v assumpta sit pro $x - \frac{1}{2}a$, si ab A versùs E sumatur A H $\infty \frac{1}{2}a$: erit, juxta Regulam, H centrum, & semi-diameter transversa (velut H F ab una, & H G ab altera parte) $\infty \frac{1}{2}b$; ita ut diameter transversa F G (quæ quidem, ob applicatam C E ad eandem perpendicularem, transversus quoque axis est,) sit ∞b. Ratio autem transversæ diametri ad parametrum, seu quadrati transversæ ad quadratum secundæ diametri erit, ut bb ad $bb - aa$. Vnde per ea, quæ Capitibus tertio & ultimo libri primi exposita sunt, quæsita Ellipsis facillimè describetur. Porrò cum quadratum semi-diametri transversæ sit $\infty \frac{1}{4}bb$, erit quadratum semi-secundæ diametri $\infty \frac{1}{4}bb - \frac{1}{4}aa$. Atqui cum F B seu G A sit $\infty \frac{1}{2}b + \frac{1}{2}a$, & B G seu A F $\infty \frac{1}{2}b - \frac{1}{2}a$, erit quoque rectangulum F B G seu G A F $\infty \frac{1}{4}bb - \frac{1}{4}aa$, nempe æquale quadrato semi-secundæ diametri, sive, ut Veteres loquebantur, æquale quadranti figuræ ad transversum axem factæ. Ideoque puncta A & B ea ipsa sunt, quæ vulgò Ellipseos Foci sive Vmbilici nuncupantur. Vnde apparet, ex præmissis rectè inferri, quæ sequuntur.

Corollarium 1.

Quæ à quolibet in Ellipsi puncto ad utrumque Vmbilicum rectæ ducuntur , simul sumptæ transverso axi æquales sunt.

Quemadmodum autem in Hyperbola superiùs demonstratum est, ductarum C A, C B differentiam transverso axi F G æquari, ita & hîc earum aggregatum eidem transverso axi æquale esse ostendetur, nempe, si non per additionem & compositionem,

[302]

as in the former case, but the proof is based on subtraction and on division [3.43]. It seems possible, however, to deliver this proof more elegantly [3.44] in the following way by introducing some changes [3.45].

Let FCG be an arbitrary ellipse with H as its center, FG as its major axis, OP as its minor axis [3.46] and A and B as its umbilici, and so that the rectangle FBG, just as GAF, equals the square on half the second diameter HO [3.47].

From point C selected at random on the curve, we draw the segments CA and CB and then we apply CE and CN ordinate-wise to each of the axes and we select M so that HE is to HM as HF is to HA, whence[1] the rectangle AHE is equal to the rectangle FHM; we also choose HQ equal to HE.

Because[2] HE^2 is to HM^2 as HF^2 is to HA^2, therefore (by conversion of a ratio [3.30]), HE^2 will be to EMQ as HF^2 is to GAF (or HO^2)[3], i.e.[4] as CN^2 (or HE^2) is to ONP and so[5] the rectangles ONP and EMQ are equal.

Because[6] HM^2 together with EMQ, i.e. together with the rectangle ONP, is equal to HE^2 and as[7] HF^2 equals the sum of the squares on HA and on HO[8] (i.e.[9] the sum of the square on CE and the rectangle ONP), $HM^2 + ONP + HF^2$ will therefore be equal to $HE^2 + HA^2 + CE^2 + ONP$.

If we subtract the rectangle ONP on both sides, then the two squares on the segments HM and HF (or HG) will remain being together equal to the three squares on the segments HE, HA (or HB) and CE.

---

[1] VI, 16.

[2] VI, 22.

[3] By hypothesis.

[4] Lib. I, Prop. 13, p. [209].

[5] V, 9.

[6] II, 5.

[7] II, 5.

[8] Because, by hypothesis, the square on HO is equal to the rectangle GAF.

[9] II, 5.

302   E L E M.   C V R V A R V M

ut ibidem factum est, sed per subductionem & divisionem argumentatio instituatur. Quod ipsum tamen, adhibitâ nonnullâ mutatione, elegantiùs quoque in hunc modum absolvi posse videtur.

Esto quælibet Ellipsis F C G, cujus centrum H, axis major F G, minor O P, atque Vmbilici A & B; adeoque rectangulum F B G ut & G A F æquale quadrato semi-secundæ diametri H O.

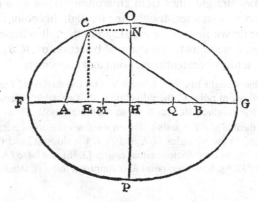

Ductis ab assumpto quolibet curvæ puncto C rectis C A, C B, ordinatim ad utrumque axem applicentur C E, C N; & fiat ut H F ad H A, ita H E ad H M, adeò ut [1] A H E rectangulo æquale sit rectangulum F H M; sumaturque H Q æqualis ipsi H E. Hinc cum sit [2] ut H F q ad H A q, ita H E q ad H M q, erit quoque per conversionem rationis ut H F q ad G A F seu [3] H O q, id est [4], ut C N q sive H E q ad O N P, ita idem H E q ad E M Q; ac proinde [5] æqualia sunt rectangula O N P & E M Q. Quocirca cum [6] H M q unà cum E M Q, id est, cum O N P rectangulo, æquale sit H E q; sitque & H F q [7] æquale quadratis rectarum H A & (H O [8] seu [9]) C E unà cum rectangulo O N P: erunt H M q + O N P + H F q æqualia H E q + H A q + C E q + O N P. Ac proinde si utrinque auferatur O N P rectangulum, remanebunt bina quadrata rectarum H M & H F seu H G simul æqualia tribus quadratis rectarum H E, H A seu H B, & C E.

Hinc

[1] per 16 sexti.
[2] per 22 sexti.
[3] ex hypothesi.
[4] per 13 primi hujus ejusque Corol. 1
[5] per 9 quinti.
[6] per 5 secundi.
[7] per 5 secundi.
[8] quippe quadr. ex H O æquale est G A F rectang. ex hypoth. [9] per 5 secundi.

[303]

Hence the following holds: if we add or subtract equals on both sides, namely twice FHM (or GHM) on one side and twice AHE (or BHE) on the other side, then[1] FM^2 will be equal to $AE^2 + CE^2$, i.e.[2] equal to AC^2 and in the same way[3] GM^2 equals $BE^2 + CE^2$, i.e.[4] BC^2. Because for this reason FM equals AC and GM equals BC, the sum of AC and BC will therefore be equal to the transverse axis FG. Which was to be demonstrated.

Corollary 2

If one draws straight lines from an arbitrary point on an ellipse to both umbilici and if one draws another straight line through this point forming equal angles with each of the drawn lines [note transl.: from the following explanation it is clear that this statement pertains to the external bisection of ACB], then this line will touch this curve at the aforementioned point and conversely.

Indeed, if the straight line ICK (drawn so that the angles ACI and BCK are equal) does not touch the ellipse at point C, then this line will, if possible, intersect the ellipse at C and at K. Next we first produce AC up to L so that the whole AL is equal to the axis FG and, therefore,[5] the added CL equals CB; then we draw the connecting segments AK, BK and LK. As, therefore, in the triangles LCK and BCK the sides LC and CK are respectively equal to BC and CK and as they enclose equal angles [3.48], the base LK will therefore also be equal to the base BK. As, however, point K is supposed to be situated on the ellipse, therefore,

---

[1] II, 7.

[2] I, 47.

[3] II, 4.

[4] I, 47.

[5] This theorem, Cor. 1, p. [301].

Lib. II. Cap. III.   303

Hinc additis ablatiſvè ab utraque æquationis parte æqualibus, nimirum F H M ſeu G H M bis ab una, & A H E ſeu B H E bis ab altera parte: erit [1] F M q æquale (A E q + C E q, id eſt [2],) A C q : [1] per 7 itemque [3] G M q æquale (B E q + C E q, id eſt [4],) B C q. Cum-
que propterea recta F M æquetur ipſi A C , & G M ipſi B C : erit ipſarum A C & B C aggregatum tranſverſo axi F G æquale. Quod demonſtrandum erat.

[1] per 7 ſecundi.
[2] per 47 primi.
[3] per 4 ſecundi.
[4] per 47 primi.

Corollarium 2.

Ductis à quolibet Ellipſeos puncto ad utrumque Vmbilicum rectis, ſi per idem illud punctum altera recta agatur , æquales cum utraque ducta angulos conſtituens, eadem curvam in dicto puncto contingit; & contra.

Si enim recta I C K ita ducta, ut æquales ſint anguli A C I, B C K, non contingat Ellipſin in C puncto , ſecet eandem, ſi fieri poteſt, in C & K. Deinde productâ A C ad L, ita ut tota A L

Fig. 1.

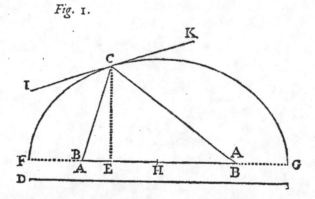

axi F G, ideoque [5] adjecta C L ipſi C B æqualis ſit , jungan-
tur A K, B K, L K. Cum igitur, in triangulis L C K, B C K latera L C, C K lateribus B C, C K, utrumque utrique, circa æquales angulos, æqualia ſint, erit quoque baſis L K baſi B K æqualis. At verò cum punctum K in Ellipſi ſupponatur, erunt,

[5] per Co-
rol. 1 hu-
jus.

per

[304]

by the preceding corollary, the segments AK and KB, i.e. the sides AK and KL, taken together, will be equal to the transverse axis FG and so also equal to the base AL, which is absurd.[1] So the straight line ICK does not intersect the ellipse, but it touches the ellipse at point C. And as no other straight line than ICK^2 can touch the ellipse at C, it is conversely clear that the straight line that touches the ellipse at C, causes the angles ACI and BCK to be equal.

CHAPTER IV

General Rule to find and to determine arbitrary plane and solid loci

[4.1]

Now, after all these preliminaries, we can infer the general Rule that all equations that can turn up and occur in the investigation of loci in the aforesaid way, have one of the following forms or can be reduced to one of them by means of the method explained before, provided that in them neither a product of two unknown quantities by itself nor a mutual product of them rises to the third degree,

---

[1] I, 20.

[2] Lib. I, Prop. 17, p. [223].

304   E L E M. C V R V A R V M

per Corollarium præcedens, rectæ A K, K B, hoc est, latera
A K, K L simul sumpta transverso axi F G, ideoque & basi A L

Fig 11.

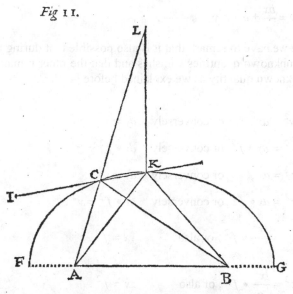

₁ *per 20* æqualia, quod est absurdum [1]. Non igitur secat recta I C K El-
primi. lipsin, sed eandem contingit in C puncto. Cumque non possit in
₂ *per 17* eodem puncto C alia recta Ellipsin contingere quàm I C K [2], ma-
primi hu- nifestum est, è contra quoque eam, quæ Ellipsin in C contingit,
jus. efficere angulos A C I, B C K æquales.

C A P V T IV.

Regula universalis inveniendi ac determinandi loca quælibet plana & solida.

Iam verò his omnibus ita præmissis, pro generali
Regula concludi potest, æquationes omnes, quæ in in-
dagatione Locorum prædicto modo obvenire atque
obtingere possunt, ita ut in iis neutra quantitatum incognitarum in se ducta, neque factum sub iisdem ad solidum

[305]

but that such a product does not exceed a square or a product of two different factors. These forms are:

1. $y = bx / a$, or, which is the same, $y = x$, as we may suppose $a = b$.

$$y = \frac{bx}{a} \pm c \text{ or } y = c - \frac{bx}{a}.$$ [4.2]

But here we have to remark that it is also possible that during the operation one of the two unknown quantities vanishes and that the other remains on its own, equal to some known quantity as we explained before [4.3].

2. $y^2 = dx$,   or conversely $dy = x^2$.

$y^2 = dx \bullet f^2$,or conversely $dy \bullet f^2 = x^2$.

$z^2 = dx$,   or conversely $dy = v^2$.

$z^2 = dx \bullet f^2$,or conversely $dy \bullet f^2 = v^2$.   [4.4]

3. $y^2 = \dfrac{lx^2}{g} \bullet f^2$, or also $yx = f^2$.

$z^2 = \dfrac{lx^2}{g} \bullet f^2$, or also $zx = f^2$.

$y^2 = \dfrac{lv^2}{g} \bullet f^2$, or also $yv = f^2$.

$z^2 = \dfrac{lv^2}{g} \bullet f^2$, or also $zv = f^2$.

Here we suppose that y and x are everywhere undetermined quantities that we conceived at the beginning, but that z is an adopted quantity composed of y plus or minus some other quantity either totally known or even combined with the other unknown quantity that we conceived at the beginning, namely x, and we suppose that v, however, has also been adopted, but in this case only consisting of x plus or minus another known quantity and in no way combined with the unknown quantity y. Or, on the other hand we suppose that v is x plus or minus some other quantity

L i b II. C a p IV.   3ᵟ5

lidum excurrat, fed aut quadratum, aut planum non
excedat, ex aliqua fequentium formularum conftare,
vel ad earundem aliquam Methodo jam explicatâ reduci poffe: nimirum,

$$1^{mò}\cdot\begin{cases} y \infty \frac{b\,x}{a}, \text{ five, quod idem eft, } y \infty x : \text{ cum fupponi} \\ \qquad \text{poffit effe } a \infty b. \\ y \infty \frac{b\,x}{a} \; 8 \; c, \text{vel} \, y \infty c - \frac{b\,x}{a}. \end{cases}$$

Signum 8
fignificat
+ *vel* —.

Sed hîc notandum, fieri etiam poffe, ut per operationem quantitatum incognitarum altera evanefcat,
alteraque fola notæ alicui quantitati æqualis remaneat,
ficut fuperiùs expofitum eft.

$$2^{dò}\cdot\begin{cases} yy \infty d\,x, \text{ aut converfim } d\,y \infty x\,x. \\ yy \infty d\,x.ff, \text{ aut converfim } d\,y.ff \infty x\,x. \\ zz \infty d\,x, \text{ aut converfim } d\,y \infty v\,v. \\ zz \infty d\,x.ff, \text{ aut converfim } d\,y.ff \infty v\,v. \end{cases}$$

$$3^{tiò}\cdot\begin{cases} yy \infty \frac{l\,xx}{g}.ff. \\ zz \infty \frac{l\,xx}{g}.ff. \\ yy \infty \frac{l\,vv}{g}.ff. \\ zz \infty \frac{l\,vv}{g}.ff. \end{cases} \quad \text{five etiam} \quad \begin{cases} y\,x \infty ff. \\ z\,x \infty ff. \\ y\,v \infty ff. \\ z\,v \infty ff. \end{cases}$$

Supponendo ubique *y* & *x* effe quantitates indeter-
minatas ac primò conceptas; at verò *z* effe quantitatem affumptam, & quæ compofita fit ex *y* 8 aliâ quâdam quantitate, vel in totum cognitâ, vel cui etiam
altera incognita primùm concepta, nimirum *x*, permixta fit; atque *v* quidem affumptam quoque effe, fed
eo cafu conftare folummodo ex *x* 8 aliâ quantitate
cognitâ, abfque ulla ipfius *y* incognitæ quantitatis per-
mixtione: aut contra *v* effe ∞ *x* 8 aliâ quâdam quan-

Pars II.   Qq   titate,

[306]

that may be combined with the unknown y; in this case z consists of y plus or minus another totally known quantity [4.5].

If the equation is similar to some of the forms included in No. 1, then the locus will be a straight line; if included in No. 2, a parabola; if included in No. 3, it will be a hyperbola, an ellipse or a circle according to differences among the signs and the angles [4.6].

To determine, however, the aforesaid loci in detail or to describe the aforesaid curves in the plane in a geometrical way one has to know that we have to presuppose some point and some line; the point serves as the initial point of one of the unknown quantities that we conceived at the beginning and we suppose that this quantity extends indefinitely along this line. Moreover, we have to presuppose some angle enclosed by the aforesaid unknown quantities at the point at which they are supposed to be linked together [4.7].

So let in the added figure, as well as in all following figures, point A be the aforesaid point and let line AB be the aforesaid line and let us suppose that the quantity x extends indefinitely from A along AB and let ABE be the angle enclosed by the quantities y and x, that are linked together at point B.

In the first case, in which the required locus is a straight line, namely when the equation is $y = x$ or $y = bx/a$, point A itself will be the initial point of the aforesaid line and to describe this line in detail we have to select a point at random on line AB, for example B.

Next we first draw a straight line through it, for example HBE, so that the angle ABE is equal to the angle that we presupposed or conceived. If we select a point on this straight line (for example D) so that AB and BD are equal or so that AB is to BD as a is to b and if we draw from A the straight line AD through point D, then this AD will be the required locus, if extended indefinitely. But if a term c is also found in the equation and if this term is provided with the plus sign, then we have to draw the segment AF from A, parallel to HBE and equal to the known c and on the same side of AB as point E, or on the opposite side if c is provided with the minus sign. If we next draw FE or FG (intersecting AB at O) parallel to AD, then FE or OG will be the required locus, if produced indefinitely [4.8].

306   ELEM. CVRVARVM

titate, cui & y incognita permixta esse possit atque eo
quidem casu z ex y $\mathcal{B}$ aliâ quantitate in totum cognitâ
constare.

Et si æquatio similis sit alicui formularum sub N° 1.
comprehensarum, erit Locus quæsitus Linea Recta;
sub N° 2. Parabola; & sub N° 3. secundùm signorum
angulorumque varietatem vel Hyperbola, vel Ellipsis,
vel Circulus.

Vt autem prædicta Loca specificè determinentur sive prædictæ Lineæ in plano Geometricè describantur, sciendum est, aliquod debere præsupponi punctum, ut & aliquam lineam à quo
exordium sumat, & per quam indefinitè se extendere intelligatur altera incognitarum quantitatum primò conceptarum; itemque angulum quendam esse præsupponendum, quem dictæ quantitates incognitæ constituant in puncto, in quo sibi invicem junctæ intelliguntur.

Sit itaque in apposita figura, ut & in sequentibus omnibus,
prædictum punctum A, dictaque linea A B, à quo, & per quam
quantitas x se indefinitè extendere concipiatur; atque angulus
A B E, quem faciunt quantitates y & x, in puncto B sibi invicem
junctæ.

Et primo quidem casu, cùm Locus quæsitus est Linea recta,
nimirum, æquatione existente $y \infty x$ vel $y \infty \frac{bx}{a}$, ipsum A punctum erit initium dictæ lineæ, atque ut eadem specificè describatur sumendum est in linea A B punctum utcunque, exempli gratiâ, B, ac per illud ductâ rectâ, velut H B E, ita ut angulus A B E
præsupposito vel concepto angulo sit æqualis, si in eadem rectâ
sumatur punctum, veluti D; ita ut A B & B D sint æquales, vel
ut A B sit ad B D, sicut a ad b, atque ex A per punctum D ducatur recta A D: erit eadem A D indefinitè extensa Locus quæsitus. At si in æquatione inveniatur quoque terminus c, ac ipse quidem signo $+$ affectus sit, ducenda est è puncto A ad eandem partem lineæ A B quàm est punctum E, aut si signo $-$ adficiatur ab
alterâ parte, recta A F ipsi H B E parallela atque æqualis c cognitæ; ductâque F E vel F G, quæ rectam A B secet in O, ipsi A D
parallelâ: erit F E vel O G indefinitè producta Locus quæsitus.
                                                          Sed

[307]

But if the equation is $y = c - (bx / a)$, we have to select point H [4.9] on the aforesaid line HBE on the opposite side of line AB from the chosen angle ABE, so that AB is to BH as a is to b.

Next we first draw AH and then we have to draw FI from the aforesaid point F on the opposite side of line AB from point H parallel to AH. Then this FI will be the required locus, if produced until [4.10] it intersects line AB [4.11].

Indeed, because both AB and BD are equal to x, or AB is to BD, on one side (of AB, transl.) and AB is to BH on the other side, as a is to b and as, therefore, BD or $BH = (bx / a)$ and as moreover AF or DE, or DG as well as HI are equal to the known c, therefore

$$BE = BD + DE = \frac{bx}{a} + c$$

and

$$BG = BD - DG = \frac{bx}{a} - c$$

and

$$BI = HI - HB = c - \frac{bx}{a}.$$

As point B has been chosen at random, the same demonstration will hold for all other points selected on line AB or on the aforesaid loci. And so it is clear that the aforesaid lines AD, FE, FG and FI are the required loci. Which was to be determined and to be demonstrated.

Lɪʙ. II. Cᴀᴘ. IV. 3o7

Sed si æquatio sit $y \infty c - \frac{bx}{a}$, in dicta linea H B E sumendum est
ab altera parte lineæ A B, quâ datus vel assumptus angulus A B E
existit, punctum H, ita ut A B ad B H sit, sicut *a* ad *b*; ductâque
A H, ex prædicto puncto F ab opposita parte lineæ A B, quâ sum-

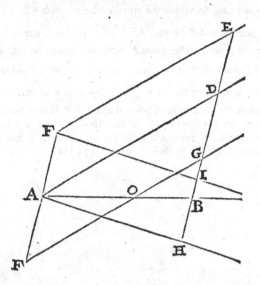

ptum est punctum H, ducenda est F I ipsi A H parallela: eritque
eadem F I producta donec cum linea A B coïncidat Locus quæsitus.

Etenim, cum tam A B quàm B D sit ∞ *x*, aut A B ad B D ab
una, ut & A B ad B H ab altera parte, sit ut *a* ad *b*; ac proinde
B D vel B H $\infty \frac{bx}{a}$: itemque, cum A F seu D E vel D G ut & H I
sint æquales *c* cognitæ: erit B E sive B D + D E $\infty \frac{bx}{a} + c$, & B G
sive B D — D G $\infty \frac{bx}{a} - c$, ac B I sive H I — H B $\infty c - \frac{bx}{a}$. Vnde cum punctum B sumptum sit utcunque, eadem erit de omni-
bus aliis, in linea A B, prædictisvè locis, assumptis punctis demonstratio: atque ita patet prædictas lineas A D, F E, F G, &
F I esse Loca quæsita. Quod determinandum, demonstrandumque erat.

Qq 2   At

[308]

But if the required locus is a parabola according to the formulas exhibited in No.2, we will distinguish the following cases [4.12], [4.13].

I.   In the first case, in which the equation is $y^2 = dx$, AB itself is the diameter of the parabola; the lines ordinate-wise applied to it enclose angles equal to the given or chosen angle ABE and point A is its vertex [4.14].

II.   In the second case, in which the equation is $y^2 = dx \bullet f^2$, the diameter remains on the same line AB and if we put $AF = (f^2 / d)$, as in the added figure, its vertex will be situated at point F. This point F, however, has to be situated on the opposite side of point A from point B if both terms dx and f^2 are provided with the plus sign. But if either the term dx or the term f^2 is provided with the minus sign, point F has to be situated on the same side of point A as point B. If then the term dx is provided with the plus sign, the parabola has to be described from A toward B, but if, in the opposite case, the term dx is provided with the minus sign, it has to be described in the opposite direction, namely from F toward A [4.15].

308   ELEM. CVRVARVM

At verò si juxta formulas sub N°. 2 exhibitas Locus quæsitus sit linea Parabolica, erit

I. Primo casu, quando æquatio est $yy \infty dx$, ipsa A B Parabolæ diameter, ad quam ordinatim applicatæ faciant angulos, dato vel assumpto angulo A B E æquales, atque ejusdem vertex A punctum.

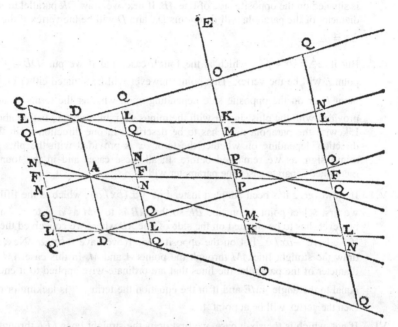

I I. Secundo casu positâ æquatione $yy \infty dx.ff$, manente diametro in eadem linea A B, sumptâque, ut in sequenti figura, A F $\infty \frac{ff}{d}$, erit ejusdem vertex in puncto F. Quod quidem punctum F, si uterque terminus tam dx quàm ff signo + sit affectus, ab altera parte puncti A, quâ est punctum B, sumendum est; sed si vel terminus dx, vel terminus ff signo — affectus sit, ab eadem parte puncti A, quâ est punctum B, sumi debet: & quidem si terminus dx signo + affectus sit, ab A versùs B Parabola describenda est; sin contra terminus dx signo — affectus fuerit, in contrariam partem, ab F nempe versùs A, describi debet.

At

[309]

But if the equation is $z^2 = dx$ or $z^2 = dx \bullet f^2$ [4.16], in which z is not the quantity we conceived at the beginning, but a quantity we adopted, then it will be adopted instead of $y \pm c$, or of $y \pm (bx/a)$, or, finally, instead of $y \pm bx/a \pm c$.

III. Now, if z has been adopted instead of $y \pm c$, which is the third case, we have to draw the segment AD through point A, parallel to BE and equal to c, so that point D is situated on the side of AB where the angle ABE is conceived if z has been adopted instead of $y - c$ and on the other hand, if z has been adopted instead of $y + c$, point D is situated on the opposite side of line AB. If next we draw DK parallel to AB, then the diameter of the parabola will lie on this DK and D will be the vertex if the equation is $z^2 = dx$.

IV. But if $z^2 = dx \bullet f^2$, which is the fourth case, and if we put $DL = f^2/d$, then point L will be the vertex. This point, however will be situated either on this side of point D or on the opposite side depending upon whether the terms dx and f^2 are provided with the plus sign or with the minus sign, like we said before about point F. Likewise the parabola itself has to be described in one direction or in the opposite direction depending on whether the term dx is provided with the plus sign or the minus sign, as we remarked before. In all these cases and in the four previously mentioned separate cases the parameter will be d.

V. If, however, z has been adopted instead of $y \pm (bx/a)$, which is the fifth case, then we first select point M on line BE so that AB is to BM as a is to b. This point M, however, has to be situated on the side of line AB where we conceived the angle ABE if we have $-bx/a$, but on the opposite side if we have $+bx/a$. Next we have to draw the straight line AM through the points A and M. In this case AM will be the diameter of the parabola; the lines that are ordinate-wise applied to it enclose angles equal to the angle AME and if in the equation the term f^2 is lacking or equals zero, then the vertex will be at point A.

VI. If not, which is the sixth case, we first draw the straight lines LFL through the points F and L [4.17] intersecting the aforementioned diameter AM or its "conjugate" [4.18] at point N. Then the vertex will be at N, situated either on this side of point A or on the opposite side, depending upon whether the terms dx and f^2 in this equation are provided with the plus sign or with the minus sign. Likewise the parabola itself has to be described in one direction or in the other, depending upon whether the term dx is provided with the plus sign or with the minus sign as we remarked before.

LIB. II. CAP. IV. 309

At si æquatio sit z z ∞ d x, vel z z ∞ d x ff, cum z non sit quantitas primò concepta sed assumpta, vel assumpta erit pro y ⅁ c, vel pro y ⅁ $\frac{b\,x}{a}$, vel denique pro y ⅁ $\frac{b\,x}{a}$ ⅁ c.

III. Et si quidem z assumpta sit pro y ⅁ c, qui sit casus tertius, ducenda est per punctum A recta A D ipsi B E parallela atque ∞ c; ita ut, si z assumpta sit pro y — c, punctum D cadat ad eandem partem lineæ A B, quam conceptus est angulus A B E: Et, si z sit assumpta pro y + c, punctum D è contra ad alteram partem lineæ A B cadat. Deinde ductâ D K ipsi A B parallelâ, erit in eadem D K Parabolæ diameter, & D vertex, si æquatio sit z z ∞ d x.

IV. Sed si sit z z ∞ d x. ff, qui sit quartus casus, sumptâ D L ∞ $\frac{ff}{d}$, erit vertex punctum L; quod quidem pro terminorum d x & ff per + vel — affectione eodem modo, ut supra de puncto F dictum est, vel citra vel ultra D punctum cadet; uti & vel in hanc vel in illam partem, prout terminus d x signo + vel — adfectus fuerit, ipsa Parabola, ut supra notatum est, describi debet: eritque omnibus & singulis prædictis quatuor casibus Parameter ∞ d.

V. Si verò z assumpta sit pro y ⅁ $\frac{b\,x}{a}$, qui casus sit quintus, sumpto in linea B E puncto M, ita ut sit A B ad B M, sicut a ad b, (quod quidem punctum M sumendum est ab eadem parte lineæ A B, quâ conceptus est angulus A B E, si habeatur — $\frac{b\,x}{a}$, sed ab altera parte, si habeatur + $\frac{b\,x}{a}$) ducenda est per puncta A & M recta A M: eritque A M eo casu Parabolæ diameter, ad quam ordinatim applicatæ faciant angulos angulo A M E æquales, & si in æquatione terminus ff deficiat aut nullus sit, erit vertex in puncto A.

VI. Sin minus, qui sit casus sextus, ductis per puncta F & L rectis L F L, quæ intersecent supra dictas diametros A M vel iis in directum adjunctas in punctis N: erit vertex in N, vel citra, vel ultra A punctum cadens, prout termini d x & ff in æquatione vel signo + vel signo — affecti fuerint; uti & vel in hanc vel in illam partem ipsa Parabola pro varia termini d x affectione, ut supra notatum est, describenda erit.

Si

[310]

If, finally, z has been adopted instead of $y \pm (bx/a) \pm c$, then we first draw AD equal to c, as we had explained just now. This point D has to be situated on this side of line AB or on the opposite side depending upon the plus sign or the minus sign with which the quantity c is provided.

VII. Then we have to draw the straight line DO from point D, parallel to that AM that lies on the same side (of AB as D, transl.) if the terms bx/a and c are provided with the same sign, which is the seventh case.

VIII. But in the other case [4.19], which is the eighth case, we have to draw the straight line DP parallel to that AM that lies on the opposite side of line AB and this DO or DP has to be chosen as a diameter; the lines that are ordinate-wise applied to it, enclose angles equal to DOE or DPE; the vertex will be point D if the term f^2 is lacking in the equation.

IX. If not [4.20], which is the ninth case, the vertex will be the common point of intersection of the diameters DO or DP with the lines LFL, namely point Q; this point is again situated on this side of D or on the opposite side, depending upon whether the terms dx and f^2 are provided with the plus sign or the minus sign. Likewise the parabola itself

310   ELEM. CVRVARVM

Si denique z aſſumpta ſit pro $y \otimes \frac{bx}{a} \otimes c$, ductâ, ut modò expoſitum fuit, A D ∞ c, ex puncto D (quod pro quantitatis c per ſignum + vel — affectione, ut ſupra, vel ab hac, vel ab illa parte lineæ A B ſumi debet) ducenda eſt recta D O ipſi

VII. A M, quæ eſt ad eandem partem, parallela, ſi termini $\frac{bx}{a}$ & c eodem ſigno ſint affecti, qui caſus ſit ſeptimus.

VIII. At ſi diverſo, qui ſit caſus octavus, ducenda eſt recta D P parallela ipſi A M, quæ eſt ab adverſa parte lineæ A B, atque eadem D O vel D P ſumenda eſt pro diametro, ad quam or-dinatim applicatæ faciant angulos angulo D O E vel D P E æquales: eritque vertex punctum D, ſi terminus ff in æqua-tione deficiat.

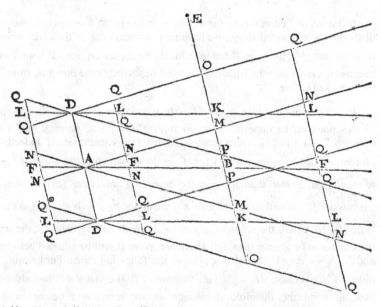

IX. Sin minùs, qui ſit caſus nonus, erit idem vertex ipſarum D O vel D P diametrorum & linearum L F L communis interſe-ctio, videlicet punctum Q, quodque iterum pro terminorum dx & ff per ſignum + vel — affectione vel citra vel ultra D punctum cadit; quemadmodum & ipſa Parabola vel verſùs hanc

[311]

has to be described in one direction or in the other depending upon the sign of the term dx, as we remarked before. In the last five cases that we already explained, the parameter will be to the known d as AB is to AM, which means that d will be to the parameter as AM is to AB [4.21].

The proof of all these statements, however, is very easy [4.22]. Indeed, let us suppose that the parabolas are described with the aforementioned diameters and parameters, passing through the indicated vertices, and let one of the lines that are ordinate-wise applied to these diameters be selected at random along a straight line OE and let us suppose that these parabolas intersect the aforementioned applied lines at point E [4.23].

In the *first case*, in which the part of the diameter AB is intercepted between A and an arbitrary line that is ordinate-wise applied to this diameter, AB is conceived as x and these applied are individually conceived as y. The parameter is d and by the nature of a parabola[1] the rectangle on the aforementioned parameter d and the segment AB equals the square on BE, we have

$$dx = y^2. \tag{1}$$

In the *second case*, where the vertex is in the point F and where there is a threefold distinction as was noted above, we have first to remark that in the cases where the equation is $y^2 = dx \pm f^2$, the point B can indefinitely be chosen on line FB from A toward B [4.24] because in these cases the parabola has to be described in the direction from A toward B, as we remarked before.

But in the case in which $y^2 = f^2 - dx$ when according to the Rule the parabola has to be described in the opposite direction from F toward A, point B may only be chosen between F and A [4.25], which is also clear from the equation itself. Indeed, because in the aforementioned equation holds $y^2 = f^2 - dx$ or, which is the same, $f^2 - y^2 = dx$, the term f^2 is greater than dx, namely exceeding this term (dx) by the quantity y^2. Therefore, f^2 / d will be greater than x if we divide by d on both sides. As according to the Rule f^2 / d equals the segment AF, and as x equals the segment AB, the segment AF will therefore also be greater than AB. Therefore, point B will be situated between the points A and F, as we stated. This also applies to the following cases. Furthermore the following holds [4.26]: because $AF = f^2 / d$, therefore FB, (i.e. $AB \pm AF$, and also $AF - AB$, if we bear in mind the threefold distinction as we mentioned before) will be equal to $x \pm (f^2 / d)$, and also to $(f^2 / d) - x$. If we multiply this equation by the parameter d the rectangle (enclosed by the parameter and AB, transl.) will be equal to $dx \pm f^2$

---

[1] Lib. I, Prop. 1, p. [162].

L ɪ ʙ. II. C ᴀ ᴘ. IV.   311

hanc vel versùs illam partem pro diversa termini dx affectione, ut supra est notatum, describenda est: Ac postremis quidem istis quinque casibus jam explicatis Parameter erit ad d
cognitam, sicut A B ad A M, hoc est, erit ut A M ad A B, ita d
ad Parametrum.

Quorum quidem omnium demonstratio perfacilis est. In-
telligantur enim Parabolæ prædictis diametris ac parametris
descriptæ, quæ per annotatos vertices transeant, sitque ordinatim ad easdem diametros applicatarum aliqua in recta O E
utcunque sumpta, & supponatur easdem Parabolas prædictam
applicatam secare in E puncto: & primo casu, cum pars diametri A B inter verticem A & quamlibet ad eandem diametrum applicatam intercepta, veluti A B, concipiatur, ut x,
ac singulæ illæ applicatæ, ut y; sitque Parameter ∞d, atque
ex natura Parabolæ [1] rectangulum sub dicta Parametro & re-   [1] per 1
1. cta A B contentum sit ∞ B E quadrato : erit $dx \infty yy$.   primi hu-
    jus.

Secundo casu, ubi vertex est in puncto F cum triplici distinctione, ut supra monitum est, notandum primò venit, in casibus, ubi æquatio est $yy \infty dx \, \beta \, ff$, punctum B in linea F B ab
A versùs B indefinitè sumi posse : cum istis casibus ab A versùs
B Parabolam describendam esse supra annotatum sit; At verò
casu, ubi æquatio est $yy \infty ff - dx$, cum juxta Regulam Parabola in contrariam partem ab F versùs A sit describenda,
punctum B non nisi inter F & A assumendum esse. id quod
etiam ex ipsa æquatione manifestum est. Quoniam enim in
prædicta æquatione $yy \infty ff - dx$ sive quod idem est $ff - yy \infty$
dx, terminus ff major est quàm dx, utpote eundem excedens
quantitate yy; idcirco quoque si utrinque divisio fiat per d,
$\frac{ff}{d}$ majus erit quàm x. Quare cum secundum Regulam $\frac{ff}{d}$ æ-
quetur rectæ A F, & $x \infty$ rectæ A B, erit similiter recta A F
major quàm A B : ideoque B punctum inter A & F puncta,
sicut dictum est, cadet. id quod ad casus quoque sequentes ap-
plicatum esto. Porrò quoniam A F est $\infty \frac{ff}{d}$, erit F B(hoc est,
observatâ triplici distinctione, ut prædictum est, A B β A F,
atque etiam A F — A B) æqualis $x \, \beta \, \frac{ff}{d}$, atque etiam $\frac{ff}{d} - x$;
eâque multiplicatâ per parametrum d, sit rectangulum $dx \, \beta \, ff$,
atque

[312]

and also to $f^2 - dx$, which is equal to the square on the applied *BE* or y^2 and so

$$y^2 = dx \pm f^2 \text{ and } y^2 = f^2 - dx \qquad (2)$$

In the *third case*, in which the vertex is at point *D* and the diameter on the straight line *DK*, the following holds: because *AD* or *BK* equals *c*, *KE* (that is *BE – BK*) equals *y – c* and *KBE* (that is *BE + BK*) equals *y + c*. As in this case *z* has been adopted instead of *y* ± *c*, *KE* and *KBE* will therefore be equal to *z*. However, *DK* or *AB* is equal to *x*, the parameter is *d* and the rectangle enclosed by the aforementioned parameter and the segment *DK*, equals the square on *KE* or *KBE*. As this square equals z^2 and as that rectangle equals *dx*, therefore

$$z^2 = dx. \qquad (3)$$

In the *fourth case*, in which the vertex is at point *L*, while the diameter is still on the straight line *DK*, the following holds: *LK* (i.e. *DK* ± *DL* and also *LD – DK* if we bear in mind the threefold distinction according to the Rule) will be equal to $x \pm (f^2 / d)$ and also equal to $(f^2 / d) - x$, because *DL* or *AF* equals f^2 / d. If we multiply this equation by the parameter *d*, then the rectangle (enclosed by the parameter and *LK*, transl.) will be equal to $dx \pm f^2$ and also equal to $f^2 - dx$, which is equal to the square on the applied *KE* or *KBE*, i.e. z^2; therefore

$$z^2 = dx \pm f^2 \text{ and } z^2 = f^2 - dx. \qquad (4)$$

In the *fifth case*, in which the vertex is at point *A* and the diameter on line *AM*, the following will hold: because *AB* (i.e. *x*) is to *BM* as *a* is to *b*, therefore *BM* = *bx / a*, and so *ME* (i.e. *BE – BM*) equals $y - (bx / a)$ and *MBE* (i.e. *BE + BM*) equals $y + (bx / a)$ [4.27]. But as in this case *z* has been adopted instead of $y \pm (bx / a)$, *ME* and *MBE* will be equal to *z*. However, as in the triangle *ABM* the angle *ABM* and the ratio of the sides *AB* and *BM* (enclosing the aforementioned angle) are known, therefore[1] the mutual ratio of the other sides of the said triangle is also known and especially the ratio of the side *AB* to *AM*, let us say as *a* to *e*. Hence the following holds: because *AB*, i.e. *x*, is to *AM* as *a* is to *e*, therefore *AM* = *ex / a*. Furthermore, as in this case according to the Rule d is to the parameter as *AM* is to *AB*, i.e. as *e* is to *a*, therefore the parameter equals *ad / e* [4.28]. If we multiply this parameter by *AM* or *ex / a*, then the rectangle (enclosed by the parameter and *AM*, transl.) will be equal to *dx*. This rectangle is equal to the square on the applied *ME* or *MBE*, i.e. equal to z^2 and so

$$z^2 = dx. \qquad (5)$$

---

[1] VI, 1.

312   E L E M. C V R V A R V M

atque etiam $ff — dx$. quod æquale est quadrato applicatæ BE
2. sive yy, ac proinde $yy \infty dx \; \text{8} \; ff$, atque $yy \infty ff — dx$.

Tertio casu, ubi vertex est in puncto D, ac diameter in recta D K, quoniam AD seu BK est ∞c: erit KE, hoc est, $BE — BK \infty y — c$; & KBE, hoc est, $BE + BK \infty y + c$. Cumque eo casu z assumpta sit pro $y \; \text{8} \; c$, erit KE & KBE ∞z. Est autem D K seu A B ∞x, parameterque ∞d, & rectangulum sub dicta Parametro & recta D K contentum ∞ quadrato ex K E vel K BE. Quare cum hoc quadratum sit ∞zz, atque
3. rectangulum illud ∞dx, erit $zz \infty dx$.

Quarto casu, ubi manente diametro in recta D K vertex est in puncto L, quoniam D L sive A F est $\infty \frac{ff}{d}$, erit L K (hoc est, observatâ triplici distinctione juxta Regulam, D K 8 D L, atque etiam L D — D K) æqualis $x \; \text{8} \; \frac{ff}{d}$, atque etiam $\frac{ff}{d} — x$. quâ multiplicatâ per Parametrum d, fit rectangulum $dx \; \text{8} \; ff$, atque etiam $ff — dx$. quod æquale est quadrato applicatæ K E vel KBE, hoc est, zz: eritque proinde $zz \infty dx \; \text{8} \; ff$, atque
4. $zz \infty ff — dx$.

Quinto casu, ubi vertex est in puncto A, diameterque in recta A M, cum sit ut a ad b, ita A B, hoc est, x, ad B M: erit BM $\infty \frac{bx}{a}$, ideoque ME, hoc est, $BE — BM \infty y — \frac{bx}{a}$, & MBE, hoc est, $BE + BM \infty y + \frac{bx}{a}$. Et quoniam eo casu z assumpta est pro $y \; \text{8} \; \frac{bx}{a}$, erit ME & MBE ∞z. At cum in triangulo A BM cognita sint & angulus A BM, & ratio laterum A B, BM, dictum angulum comprehendentium, nota quoque est ratio reliquorum dicti trianguli laterum ad invicem, atque in specie etiam lateris A B ad A M, quæ sit ut a ad e. Ac proinde cum sit ut a ad e, ita A B, h.e., x ad A M: erit A M $\infty \frac{ex}{a}$. Cumque porrò juxta Regulam eo casu sit ut A M ad A B, hoc est, ut e ad a, ita d ad Parametrum : erit Parameter $\infty \frac{ad}{e}$. Quâ multiplicatâ per A M seu $\frac{ex}{a}$ fiet rectangulum ∞dx. Quod æquale est quadrato applicatæ M E vel MBE, hoc est, zz; ac proinde
5. est $zz \infty dx$.

' per 6
sexti.

Sexto

[313]

In the *sixth* case, in which the vertex is at point N [4.29] and the diameter on the straight line NM, the following holds: because AF is to AN as AB is to AM, which means that f^2/d is to AN as a is to e, therefore $AN = ef^2/ad$ and NM (i.e. $AM \pm AN$ and also $NA - AM$, if we bear in mind the threefold distinction according to the Rule) is equal to $(ex/a) \pm (ef^2/ad)$ and also equal to $(ef^2/ad) - (ex/a)$.

If we multiply this by the parameter ad/e, then the rectangle (enclosed by the parameter and NM, transl.) equals $dx \pm f^2$ and also $f^2 - dx$.

Because this is equal to the square on the applied ME or MBE [4.27], i.e. z^2, therefore $z^2 = dx \pm f^2$ and $z^2 = f^2 - dx$. 　　　　　　　　　　　　　　　　　　　(6)

In the *seventh* case, in which the vertex is at point D and the diameter on the straight line DO, the following holds: because AD or MO equals c, therefore OE (or $BE - BM - MO$) equals $y - (bx/a) - c$ and OBE (or $BE + BM + MO$) equals $y + (bx/a) + c$ [4.30]. As in this case z has been adopted in stead of $y - (bx/a) - cc$ or in stead of $y + (bx/a) + c$, therefore OE and OBE equal z (the text gives erroneously d in stead of c, transl.).

LIB. II. CAP. IV.

Sexto casu, ubi vertex est in puncto N, & diameter in recta N M, quoniam est ut A B ad A M, ita A F ad A N, hoc est, ut a ad e, ita $\frac{ff}{d}$ ad A N : erit A N $\infty \frac{eff}{ad}$, & N M (hoc est, observatâ juxta Regulam triplici distinctione, A M ꝗ A N, atque etiam N A — A M) æqualis $\frac{ex}{a}$ ꝗ $\frac{eff}{ad}$, atque etiam $\frac{eff}{ad}$ — $\frac{ex}{a}$. Quâ multiplicatâ per Parametrum $\frac{ad}{e}$, fit rectangu-

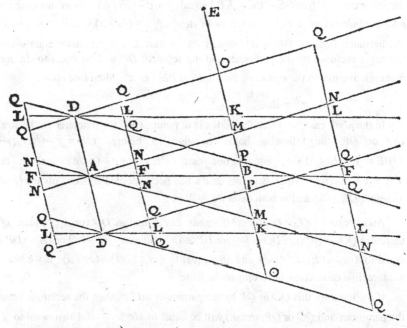

lum dx ꝗ ff, atque etiam $ff — dx$. Quod cum æquale sit quadrato applicatæ M E vel M B E, hoc est, zz : erit 6. $zz \infty dx$ ꝗ ff, atque $zz \infty ff — dx$.

Septimo casu, ubi vertex est in puncto D, & diameter in recta D O, quoniam A D seu M O est ∞r, erit O E (sive B E — B M — M O) $\infty y — \frac{bx}{a} — c$, & O B E (sive B E + B M + M O) $\infty y + \frac{bx}{a} + c$. Cumque eo casu z assumpta sit pro $y — \frac{bx}{a} — d$, vel pro $y + \frac{bx}{a} + d$: erit O E & O B E ∞z. Porrò cum D O

[314]

Furthermore, as *DO* or *AM* equals ex/a, and as the parameter of the conic section (sic) equals ad/e, the rectangle enclosed by the parameter and the segment *DO*, will therefore be equal to dx. As this same rectangle equals the square on the applied *OE* or *OBE* [4.30], i.e. z^2, therefore

$$z^2 = dx \qquad (7)$$

In the *eighth* case, in which the diameter is still on the straight line *DP*, while the vertex is still on point *D*, the following holds: because *AD* or *BK* equals *c* and $KP = bx/a$, therefore in one case *PE* (or *BE* − *BK* + *KP*) equals $y - c + (bx/a)$ and in the other case *PE* (or *BE* − *BK* − *KP*) equals $y + c - (bx/a)$. As in this case *z* has been adopted in stead of $y + (bx/a) - c$ or in stead of $y - (bx/a) + c$, in both cases *PE* = *z*. Furthermore, because *DP* or *AM* equals ex/a and as the parameter equals ad/e, the rectangle enclosed by the parameter and the segment *DP* will be equal to dx, and as this rectangle is equal to the square on both applied *PE*, i.e. z^2, therefore also

$$z^2 = dx. \qquad (8)$$

In the *ninth* case, in which the vertex is at point *Q* and the diameter on the straight line *QO* or *QP*, the following holds: because, as before, $OE = y - (bx/a) - c$ and $OBE = y + (bx/a) + c$, but in one case $PE = y - c + (bx/a)$ and in the other $PE = y + c - (bx/a)$ and as in this case *z* has been adopted in stead of $y \pm (bx/a) \pm c$, therefore *OE*, *OBE* and in both cases *PE*, will be *z*.

And because *DO* or *DP* (or *AM*) equals ex/a and as *DQ* (or *AN*) equals ef^2/ad, therefore *QO* or *QP* (i.e. *DO* ± *DQ* or *DP* ± *DQ* and also *QD* − *DO* or *QD* − *DP*) will be equal to $(ex/a) \pm (ef^2/ad)$ and also equal to $(ef^2/ad) - (ex/a)$, if we bear in mind the threefold distinction, according to the Rule.

If we multiply this *QO* or *QP* by the parameter ad/e, then the rectangle (enclosed by the parameter and *QO* or *QP*, transl.) will be equal to $dx \pm f^2$ and also equal to $f^2 - dx$. This rectangle, however, is equal to the square on the applied *OE*, *OBE* or on either *PE*, i.e. equal to z^2, therefore also $z^2 = dx \pm f^2$ and $z^2 = f^2 - dx$. (9)

These, however, are all the statements which had to be demonstrated here. Regarding, however, the equations that correspond to the converse of the nine cases above, we remark the following: to describe the parabolas that are the required loci with the same supposition as before,

feu A M fit ∞ $\frac{e\,x}{a}$, Parameterque fectionis ∞ $\frac{a\,d}{e}$, erit rectangulum fub Parametro & recta D O contentum ∞ $d\,x$. Cumque idem illud rectangulum æquetur quadrato applicatæ O E vel

7. O B E , id eft , $z\,z$: erit $z\,z$ ∞ $d\,x$.

Octavo cafu, ubi, manente vertice in puncto D , diameter eft in recta D P, quoniam A D feu B K eft ∞ c, & K P ∞ $\frac{b\,x}{a}$, erit P E una (five B E — B K + K P) ∞ $y - c + \frac{b\,x}{a}$, & P E altera (five B E + B K — K P) ∞ $y + c - \frac{b\,x}{a}$. Cumque eo cafu z affumpta fit pro $y + \frac{b\,x}{a} - c$ vel pro $y - \frac{b\,x}{a} + c$: erit utraque P E ∞ z. Porrò cum D P feu A M fit ∞ $\frac{e\,x}{a}$ ac Parameter ∞ $\frac{a\,d}{e}$, erit rectangulum fub Parametro & recta D P contentum ∞ $d\,x$. Cumque idem rectangulum æquale fit quadrato utriufque ap-

8. plicatæ P E, hoc eft, $z\,z$: erit quoque $z\,z$ ∞ $d\,x$.

Nono cafu, ubi vertex eft in puncto Q, & diameter in recta Q O vel Q P, quoniam, ut fupra, O E eft ∞ $y - \frac{b\,x}{a} - c$, atque O B E ∞ $y + \frac{b\,x}{a} + c$; at verò P E una ∞ $y - c + \frac{b\,x}{a}$, ac P E altera ∞ $y + c - \frac{b\,x}{a}$, fitque eo cafu z affumpta pro $y \,\oplus\, \frac{b\,x}{a} \,\oplus\, c$: erit O E, O B E, atque utraque P E ∞ z. Et cum D O aut D P feu A M fit ∞ $\frac{e\,x}{a}$, atque D Q feu A N ∞ $\frac{e\,ff}{a\,d}$: erit Q O vel Q P (hoc eft, obfervatâ juxta Regulam triplici diftinctione, D O vel D P $\,\oplus\,$ D Q, atque etiam Q D — D O vel D P) æqualis $\frac{e\,x}{a} \,\oplus\, \frac{e\,ff}{a\,d}$, atque etiam $\frac{e\,ff}{a\,d} - \frac{e\,x}{a}$. Vnde fi eadem Q O vel Q P multiplicetur per Parametrum ∞ $\frac{a\,d}{e}$, erit rectangulum ∞ $d\,x$ $\,\oplus\, ff$, atque etiam $ff - d\,x$. Quod quidem rectangulum cum æquale fit quadrato applicatæ O E, O B E, aut utriufque P E,

9. hoc eft, $z\,z$: erit quoque $z\,z$ ∞ $d\,x$ $\,\oplus\, ff$, atque $z\,z$ ∞ $ff - d\,x$. Quæ quidem omnia funt, quæ hîc demonftranda erant.

Quod' autem ad æquationes fuperioribus novem cafibus converfim correfpondentes fpectat, ut lineæ Parabolicæ defcribantur, quæ fint Loca quæfita : pofitis iifdem, ut fupra, per

pun-

[315]

the straight line *AC* has to be drawn through the point *A* parallel to *BE* and then *AC* is to be considered everywhere in the same way *AB* has been considered in the figure above. Furthermore, we first have to select a point at random in this *AC*, say *C*, and to draw a line through it, parallel to *AB*, say *OCE*, thereafter this *OCE* has in the same way everywhere to be considered as the straight line *OBE* was considered in the previous figure, [of course with no other change] . For instance:

I.   If the equation is $dy = x^2$, then *AC* will be the diameter, *A* the vertex and *d* the parameter. Indeed, as *AC* or *BE* is considered as *y* and *CE* or *AB* as *x* and as the rectangle enclosed by the parameter and *AC*, i.e. *dy*, equals the square on the segment *CE* or *AB*, i.e. x^2, therefore $dy = x^2$, as required.

II.   If the equation is $dy \bullet f^2 = x^2$ and if we put $AF = f^2 / d$, then *F* will be the vertex, while the diameter still lies on the straight line *FC* and the parameter is still *d* [4.31].

Lib. II. Cap. IV.   315

punctum A ducenda est recta A C ipsi B E parallela, ac deinde ipsa A C, ubique consideranda, ut considerata fuit recta
A B in superiori figura. Porrò sumpto in eadem A C puncto
utcunque, veluti C, atque per id ductâ rectâ ipsi A B parallel-
lâ, velut O C E, erit similiter hæc O C E ubique consideranda, sicut considerata fuit recta O B E in præcedenti figura,
nullâ scilicet aliâ mutatione adhibitâ. Exempli gratiâ: Si æ-
I. quatio sit $dy \infty xx$, erit A C diameter, A vertex, & Parame-

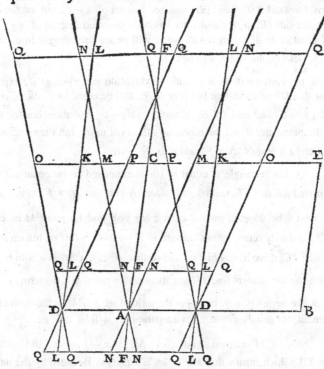

ter ∞d. Cum enim A C seu B E sit concepta ut y, & C E seu
A B ut x, rectangulumque sub Parametro & A C contentum,
hoc est, dy, æquetur quadrato rectæ C E seu A B, hoc est, xx:
erit, ut petitur, $dy \infty xx$.

II.   Si æquatio sit $dy.ff \infty xx$, sumptâ A F $\infty \frac{ff}{d}$, erit F vertex,
manente diametro in rectâ F C, atque Parametro ∞d. Est
enim

[316]

Indeed, on the basis of the threefold distinction we have according to the Rule $FC = y \pm (f^2 / d)$ and also $FC = (f^2 / d) - y$; therefore the rectangle, enclosed by the parameter and this FC, equals $dy \pm f^2$ and also $f^2 - dy$. However, as this rectangle is equal to the square on the applied CE, i.e. x^2, therefore $dy \bullet f^2 = x^2$, as is required.

III. If the equation is $dy = v^2$ or $dy \bullet f^2 = v^2$ and if v in the first place has been adopted instead of $x \pm c$, then we have to take $AD = c$ where we have to select D in the direction from A toward B if v has been assumed instead of $x - c$, but, on the other hand, on the opposite side (from A, transl.) if v has been assumed instead of $x + c$. If next we draw DK parallel to AC, then the diameter will be on the straight line DK, and if the term f^2 is lacking, the vertex will be at D.

IV. If not, the vertex will be at L, with the threefold variation, as we explained before. It is clear that DB or DAB, that is KE or KCE, will be equal to v, DK equal to y, LK equal to $y \pm (f^2 / d)$ and also LK equal to $(f^2 / d) - y$; and therefore the rectangle enclosed by the parameter d and the aforesaid DK equal to dy, but the rectangle that is enclosed by d and LK equals $dy \pm f^2$ and also $f^2 - dy$.

As this rectangle is equal to (or is supposed to be equal to) the square on the applied KE or KCE, i.e. v^2, therefore $dy = v^2$ or $dy \bullet f^2 = v^2$, as is required.

V. Let next v be adopted instead of $x \pm (by / a)$ and let point M be chosen on the line OCE, in the direction from C toward E, if we have $-by / a$, but on the opposite side of the line AC, if we have $+by / a$, and so that AC is to CM, as a is to b. Then the diameter will be on the straight line AM and its vertex at point A if the term f^2 is lacking.

VI. If not, the vertex is at N. If we put the ratio of AC to AM as the ratio of a to e, so that the segment AM equals ey / a, then the parameter will be ad / e.

Now $CM = by / a$ and also $ME = x - (by / a)$ and $MCE = x + (by / a)$ [4.32], which means that ME or MCE equals v. Because by the nature of a parabola the rectangle enclosed by the aforesaid parameter and the segment AM equals the square on ME or MCE, therefore $dy = v^2$. Because $NA = f^2 e / da$, we have $NM = (ey / a) \pm (f^2 e / da)$

3.16   ELEM. CVRVARVM

enim pro triplici juxta Regulam distinctione F C ∞ y β $\frac{ff}{d}$,

atque etiam $\frac{ff}{d}$ — y: ac proinde rectangulum sub Parametro
ac eadem F C contentum ∞ dy β ff, atque etiam ff — dy.
Quod quidem rectangulum cum æquale sit quadrato appli-
catæ C E, hoc est, xx: erit, ut petitur, $dy. ff \infty xx$.

III. Si æquatio sit $dy \infty vv$, vel $dy. ff \infty vv$, atque v primùm
assumpta sit pro x β c, factâ A D ∞ c, sumptoque puncto D ab
A versùs B, si v sit assumpta pro x — c; at contra ab altera
parte, si v assumpta fuerit pro x + c, erit, ductâ D K ipsi A C
parallelâ, diameter in recta D K. Et si terminus ff deficiat,

IV. erit vertex in D; sin secus in L, cum triplici variatione, ut
supra expositum est. Et patet, D B sive D A B, hoc est, K E
sive K C E fore v, D K ∞ y, atque L K ∞ y β $\frac{ff}{d}$, atque etiam

$\frac{ff}{d}$ — y: ac proinde rectangulum sub Parametro d dictaque
D K comprehensum ∞ dy; at verò id quod sub d & L K com-
prehenditur ∞ dy β ff, atque etiam ff — dy. Quod quidem
rectangulum cum æquale sit aut supponatur quadrato appli-
catæ K E sive K C E, hoc est, vv: erit, ut petitur, $dy \infty vv$, vel
$dy. ff \infty vv$.

V. Sit deinde v assumpta pro x β $\frac{by}{a}$, sumptoque in linea O C E
puncto M à C versùs E, si habeatur — $\frac{by}{a}$; at ab altera parte
lineæ A C, si habeatur + $\frac{by}{a}$, ita ut A C sit ad C M, sicut a
ad. b: erit in recta A M diameter, ejusque vertex in

VI. puncto A, si terminus ff deficiat; sin minus, in N. Et positâ
ratione A C ad A M, ut a ad e, ac proinde recta A M ∞ $\frac{ey}{a}$,
erit Parameter ∞ $\frac{ad}{c}$. Est enim recta C M ∞ $\frac{by}{a}$, ac proinde
M E ∞ y — $\frac{by}{a}$, atque M C E ∞ y + $\frac{by}{a}$, id est, M E vel
M C E ∞ v. Quoniam ergo ex natura Paraboles rectangulum
sub dicta Parametro & recta A M contentum ∞ quadrato ex
M E vel M C E, erit, $dy \infty vv$.

Porrò cum N A sit ∞ $\frac{ffe}{da}$, erit N M ∞ $\frac{ey}{a}$ β $\frac{ffe}{da}$, atque
etiam

[317]

and also $(f^2 e / da) - (ey / a)$. Therefore, the rectangle enclosed door de parameter and the segment NM equals $dy \pm f^2$ and also to $f^2 - dy$. As, however, this rectangle equals the square op ME or MCE, that is v^2, therefore also $dy \bullet f^2 = v^2$.

VII, VIII, IX. Let, finally, v be adopted instead of $x \pm (by / a) \pm c$. Then, with the same suppositions as before, the diameter wil be on DO or on DP and the vertex at D, if the term f^2 is lacking; if not, the vertex will be at Q. If we put the ratio of DK to DO, as well as the ratio of DK to DP at the ratio a to e, so that the segment DO, as well as DP, equals ey / a, then the parameter equals ad / e. Now $OE = x - (by / a) - c$ and

LIB. II. CAP. IV.   317

etiam $\frac{ffe}{da} - \frac{ey}{a}$: ideoque rectangulum sub Parametro
& recta N M contentum ∞ dy $\mathcal{8}$ ff, atque etiam $ff - dy$.
Quod quidem rectangulum cum sit ∞ quadrato ex M E
vel M C E, hoc est, vv; erit quoque $dy. ff \infty vv$.

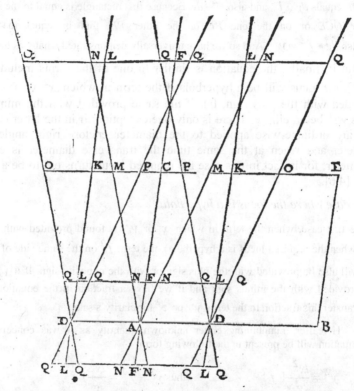

Sit denique v assumpta pro x $\mathcal{8}$ $\frac{by}{a}$ $\mathcal{8}$ c: eritque, sup-
VII.VIII. positis iisdem quæ supra, diameter in D O, vel in D P;
IX.   &, si terminus ff deficiat vertex in D; sin minus, in Q.
Et positâ ratione D K ad D O, ut & D K ad D P, sicut
a ad e, ac proinde rectâ D O, ut & D P ∞ $\frac{ey}{a}$; erit pa-
rameter ∞ $\frac{ad}{e}$. Est enim O E ∞ $x - \frac{by}{a} - c$, atque

R r 3   O C E

[318]

$OCE = x + (by/a) + c$ and $PE = x - (by/a) + c$ in   one   case   and
$PE = x + (by/a) - c$ in the other case, which means that OE, OCE and PE (in both
cases) equal v. QO or QP equals $(ey/a) \pm (f^2 e/da)$ (as NM before) and
also $(f^2 e/da) - (ey/a)$ and so the rectangle enclosed by the parameter and QO or
QP equals $dy \pm f^2$ and also $f^2 - dy$. Because this rectangle is equal to the square on OE
or OCE or on the one PE or the other PE, that is, equal to v^2, therefore
also $dy \bullet f^2 = v^2$. And so we have universally demonstrated what was here proposed.

But if, finally, the equation is similar to one of the forms included in No. 3
[4.33], the locus will be a hyperbola if the term in which x^2 or v^2 is found, is
provided with the plus sign; but if the same provided with the minus sign, the
locus will be an ellipse. There is only one exception: if in the latter case the lines
that are ordinate-wise applied to the diameter enclose right angles with this
diameter and when at the same time the transverse diameter is equal to the
parameter; for in fact in this case the required locus turns out to be a circle, as is
clear [4.6].

First case where the locus is a hyperbola

In the first case, when the term in which x^2 or v^2 is found provided with the plus sign
(when the required locus is a hyperbola), the term f^2 on the same side of the equation,
will also be provided with the plus sign or with the opposite sign. If it (f^2, transl.) is
provided with the minus sign and if we have a fraction in the equation, then let us
transfer this fraction to the term y^2 or z^2 for clarity's sake.

Hereafter, maintaining either unknown quantity as it was conceived first, the
equation will be present in the following form:

$$y^2 = \frac{lx^2}{g} + f^2 \qquad \qquad \text{(thus } y^2 - f^2 = \frac{lx^2}{g}\text{)}$$

or

$$\frac{ly^2}{g} = x^2 - f^2.$$

In the first case, as in the following figure, in which the term f^2 is at the same side of the
equation as the term containing x^2, and is provided with the plus sign, the diameter of
the hyperbola has to be described on the segment AX drawn through point A and
parallel to BE, which is given in position. But in the opposite case – that is if the term
f^2 is provided with the minus sign (as in the second case) – the diameter will be on
the straight line AB, of which the position is given and which is indeterminably
conceived as x.

318 E·L E M. C V R V A R V M

$O\,C\,E \propto x + \frac{by}{a} + c$;itemque $P\,E$ una $\propto x - \frac{by}{a} + c$, ac $P\,E$ alte-

ra $\propto x + \frac{by}{a} - c$, hoc eſt, $O\,E$, $O\,C\,E$, & $P\,E$ una vel altera

erit $\propto v$. Eſtque $Q\,O$ vel $Q\,P$ (ſicut ſupra $N\,M$) $\propto \frac{ey}{a} \ominus \frac{ffe}{da}$, at-

que etiam $\frac{ffe}{da} - \frac{ey}{a}$: ac proinde rectangulum ſub Parametro &

$Q\,O$ vel $Q\,P \propto dy \ominus ff$, atque etiam $ff - dy$. Quare cum idem

rectangulum æquale ſit quadrato ex $O\,E$ vel $O\,C\,E$, aut ex una

alteravè $P\,E$, id eſt, $v\,v$: erit quoque $dy. ff \propto v\,v$.

Atque ita demonſtratum eſt generaliter, quod hoc loco propoſitum fuit.

At ſi denique æquatio ſimilis ſit alicui formularum
ſub N° 3 comprehenſarum, erit Locus quæſitus, ſi terminus in quo invenitur $x\,x$ vel $v\,v$ ſigno $+$ ſit affectus,
Hyperbola; ſin idem terminus ſigno $-$ affectus ſit, Ellipſis: excepto tantùm, cùm poſteriori caſu ordinatim
ad diametrum applicatæ cum ea rectos angulos faciunt, & ſimul tranſverſa diameter parametro eſt æ-
qualis: quippe eo caſu, ut patet, quæſitus Locus Circulus exiſtit.

Et primo quidem caſu, cùm nempe terminus in quo $x\,x$ vel $v\,v$
ſigno $+$ affectus reperitur, ac proinde Locus quæſitus eſt Hyperbola, erit quoque terminus ff cum illo ab eadem æquationis parte conſtitutus vel ſigno $+$ affectus, vel contra; & ſi ſigno $-$ affectus ſit, atque in æquatione habeatur fractio, ipſa majoris per-
ſpicuitatis gràtiâ in terminum yy vel zz rejiciatur. Quo facto,
remanente utrâque quantitate incognitâ primùm conceptâ, ſe-

Caſus 1ⁿⁱᵘˢ, cùm Locus eſt Hyperbo-la. quenti formâ ſe exhibebit æquatio : $yy \propto \frac{lxx}{g} + ff$, (id eſt,

$yy - ff \propto \frac{lxx}{g}$) aut $\frac{lyy}{g} \propto xx - ff$: eritque, ut in ſequenti figura,
caſu primo, nempe ſi terminus ff cum termino in quo $x\,x$ unam
æquationis partem conſtituens ſigno $+$ affectus ſit, diameter Hyperbolæ deſcribendæ in recta $A\,X$, quæ ducitur per punctum A
poſitione datæ $B\,E$ parallela. Sin contra, hoc eſt, ſi terminus ff
ſigno $-$ affectus ſit, uti caſu ſecundo, erit diameter in data poſitione recta $A\,B$, quæ indeterminatè pro x concipitur; ita ut ad
eaſdem

[319]

And, the lines that are ordinate-wise applied to these diameters enclose angles equal to the given or chosen angle ABE. In either case the center of the hyperbola will be at point A and half the latus transversum will be f; which is represented on the aforementioned diameters by the segments AC or AF respectively [4.34].

Furthermore, if $l = g$ or, which is the same, if the term x^2 or y^2 is free of a fraction,

L I B. II. C A P. IV.   319

eafdem diametros ordinatim applicatæ faciant angulos, dato vel
affumpto angulo A B E æquales : eritque cafu utroque centrum
Hyperboles in puncto A, & femi-latus tranfverfum ∞ f, quod in

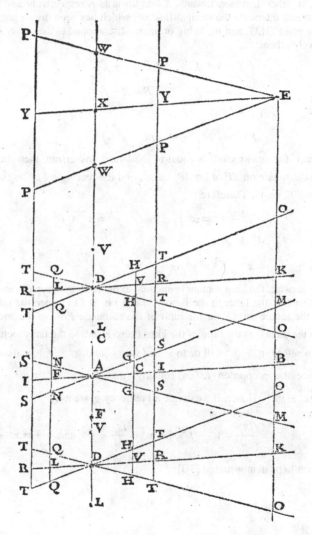

dictis diametris refpective per lineas A C vel A F exprimatur.
Porrò fi l fit ∞ g, vel, quod idem eft, fi termino x x vel y y nulla adhæreat

[320]

then the latus transversum and the latus rectum will be equal to each other. But if we suppose l and g to be unequal, then the ratio of the latus transversum and the latus rectum is equal to that of l to g [4.6].

Indeed, if we understand that the aforementioned hyperbola has been described through point C on either diameter, towards X and towards B respectively and if we suppose that this hyperbola intersects the straight line XE, which has been drawn parallel to AB, as well BE at the point E (XE and BE being ordinate-wise applied to the aforementioned diameters respectively), then:

$$FX = y + f, \qquad FB = x + f,$$
$$CX = y - f, \qquad CB = x - f \qquad\qquad [4.35]$$

and so

$$FXC = y^2 - f^2 \text{ and } FBC = x^2 - f^2.$$

However, if the latus rectum is equal to the latus transversum, then the rectangle FXC[1] is equal to the square on XE or on AB, i.e. x^2, and the rectangle FBC is likewise equal to the square on BE, i.e. y^2. Therefore

$$y^2 - f^2 = x^2 \text{ , i.e.} \qquad y^2 = x^2 + f^2$$

and also

$$x^2 - f^2 = y^2 \text{ or } y^2 = x^2 - f^2.$$

But if, otherwise, the latus rectum is unequal to the latus transversum and if the ratio of one to the other (i.e. the latter to the former, transl.) is as l to g then the ratio of the rectangle FXC to the square on XE, or the ratio of the rectangle FBC to the square on BE will be likewise the same[22] as the ratio of the latus transversum to the latus rectum, i.e. the same as l to g. Therefore $y^2 - f^2$ will be to x^2 as l is to g, and $x^2 - f^2$ is also to y^2 as l is to g. If we reduce the proportion to an equality, this means $lx^2 = gy^2 - gf^2$, as well as $ly^2 = gx^2 - gf^2$. Hence, if we divide all terms by g, we have

$$\frac{lx^2}{g} = y^2 - f^2 \text{ , i.e. } y^2 = \frac{lx^2}{g} + f^2 \text{ and } \frac{ly^2}{g} = x^2 - f^2.$$

Which was to be demonstrated [3.3].

---

[1] Prop. 10, Lib. I, p. [196]

[2] Prop. 10, Lib. I, p. [196]

hæreat fractio, erunt latera transversum & rectum sibi invicem
æqualia.' At verò positis *l* & *g* inæqualibus, erit ratio lateris transversi ad rectum ut *l* ad *g*.

Si enim descripta intelligatur prædicta Hyperbola per punctum C in utraque diametro versùs X & versùs B respectivè; supponaturque eandem secare rectam X E, quæ ducta sit ipsi A B æ-
quidiitans, ut & ipsam B E, ad dictas diametros respectivè ordinatim applicatas, in puncto E: erit

$$FX \infty y + f, \qquad\qquad FB \infty x + f,$$
$$CX \infty y - f, \qquad\qquad CB \infty x - f;$$

ideoque rectangulum

$$FXC \infty yy - ff, \quad \& \quad FBC \infty xx - ff.$$

Cum autem latere recto ipsi transverso æquali existente rectangulum F X C [1] sit ∞ quadrato ex X E seu A B, hoc est, xx;
itemque rectangulum F B C sit ∞ quadrato ex B E, hoc est, yy:
erit $yy - ff \infty xx$, hoc est, $yy \infty xx + ff$ itemque
$xx - ff \infty yy$, sive $yy \infty xx - ff$.

[1] per 10 primi hujus.

Sed cum secus recto latere ipsi transverso inæquali existente
unius ad alterum ratio sit, ut *l* ad *g*; similiterque etiam ratio rectanguli F X C ad quadratum X E, aut rectanguli F B C ad quadratum B E eadem sit [2], quæ transversi lateris ad rectum, hoc
est, eadem quæ *l* ad *g* : erit ut *l* ad *g*, ita $yy - ff$ ad xx; itemque
ut *l* ad *g*, ita $xx - ff$ ad yy, hoc est, reductâ proportione ad æ-
qualitatem, erit $lxx \infty gyy - gff$, ut & $lyy \infty gxx - gff$. unde
divisis omnibus per *g*, sit $\frac{lxx}{g} \infty yy - ff$, hoc est, $yy \infty \frac{lxx}{g} + ff$;
& $\frac{lyy}{g} \infty xx - ff$. Quod demonstrandum erat.

[2] per 10 primi hujus.

Casus 2ᵈᵘˢ, cùm Locus est Hyperbola.

At si quantitatum incognitarum primò conceptarum unâ ex
æquatione sublatâ, aliâque in ejusdem locum juxta Regulam as-
sumptâ, æquatio sit $zz \infty \frac{lxx}{g} + ff$ (id est, $zz - ff \infty \frac{lxx}{g}$), vel
$\frac{lxx}{g} \infty xx - ff$: aut *z* assumpta erit pro $y \, \beta \, c$, vel pro $y \, \beta \, \frac{bx}{a}$, aut

§. 1. pro $y \, \beta \, \frac{bx}{a} \, \beta \, c$. Et quidem primò si *z* assumpta sit pro $y \, \beta \, c$,
ducenda est per punctum A recta A D ipsi B E parallela & ∞c;
ita ut, si *z* fuerit assumpta pro $y - c$, prædictum punctum D cadat
ab eadem parte lineæ A B, quâ datus vel conceptus est angulus
A B E. Sin contra *z* fuerit assumpta pro $y + c$, idem illud punctum

[320, cont.]

Second case where the locus is a hyperbola

But if one of the unknown quantities that we conceived at first has been taken away from the equation and another has been adopted in its place according to the Rule, and if the equation is

$$z^2 = \frac{lx^2}{g} + f^2 \text{, i.e. } z^2 - f^2 = \frac{lx^2}{g} \text{, or } \frac{lz^2}{g} = x^2 - f^2 \text{,}$$

then z will be adopted instead of $y \pm c$ or instead of $y \pm (bx/a)$ or instead of $y \pm (bx/a) \pm c$.

§1. Indeed, in the first case, in which z has been adopted instead of $y \pm c$, we have to draw the segment AD through A parallel to BE and equal to c, in such a way that if z has been adopted instead of $y - c$, then the aforementioned point D falls on the same side of line AB as where the angle ABE has been given or chosen. If conversely z has been adopted instead of $y + c$,

[321]

then the same point D has to be found on the opposite side of line AB. Next we first draw the straight line DK through D parallel to AB, intersecting the straight line BE at point K (produced, if necessary). Then the diameter of the hyperbola to be described will be situated on the straight line DX if the term f^2 is provided with the plus sign. But conversely, i.e. if the term f^2 is provided with the minus sign, the diameter will be situated on the aforementioned straight line DK;

L I B. II. C A P. IV.   321

&um D reperiatur ab altera parte lineæ A B. Deinde per punctum
D ductâ rectâ D K ipsi A B parallelâ, quæ secet rectam B E productam, si opùs fuerit, in puncto K : erit describendæ Hyperbolæ

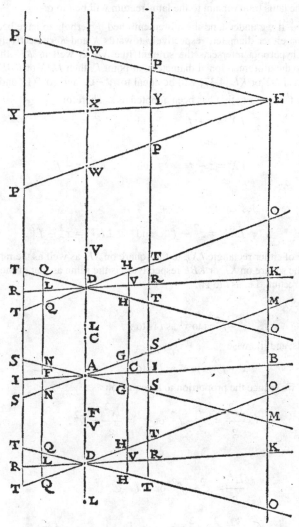

diameter, si terminus *ff* signo + affectus sit, in recta D X. sin
contra, hoc est, si terminus *ff* signo — affectus sit, in prædicta

Pars II.   S s   recta

[322]

the lines that are ordinate-wise applied to their diameters have to enclose angles that are equal to the given or chosen angle *ABE* (or *DKE*) or *DXE*.

In either case *D* will be the center and half the latus transversum equals *f*, which is represented by *DV* or *DL* respectively on the aforementioned diameters. Furthermore, the ratio of the latus transversum to the latus rectum will be *l* to *g*.

Indeed, if we understand the aforementioned hyperbola to have been described through point *V* on either diameter, respectively towards *X* and *K* respectively and if we suppose that this hyperbola intersects the straight line *XE* as well as *KE*, that are ordinate-wise applied to the aforementioned diameters, at point *E*, then *DAX* or *KBE* [4.36] will be equal to *y* + *c* and *DX* or *KE* [4.37] will be equal to *y* - *c*, and so *DAX* and *KBE* or *DX* and *KE* is exactly the quantity that has been adopted as *z*. Therefore

$$LX = z + f \qquad \text{and} \qquad LK = x + f$$

and

$$VX = z - f \qquad \text{and} \qquad VK = x - f .$$

Hence the rectangles satisfy

$$LXV = z^2 - f^2 \quad \text{and} \quad LKV = x^2 - f^2 .$$

The ratio of either rectangle *LXV* to the square on *XE*, as well as the ratio of either rectangle *LKV* to the square on *KE* or *KBE* respectively is the same as that of the latus transversum to the latus rectum, i.e. as *l* to *g*.

Therefore:

$$z^2 - f^2 \text{ is to } x^2 \text{ as } l \text{ is to } g$$

also holds, and likewise

$$x^2 - f^2 \text{ is to } z^2 \text{ as } l \text{ is to } g,$$

that is, if we reduce the proportion to an equation,

$$\frac{lx^2}{g} = z^2 - f^2 \text{ or } z^2 = \frac{lx^2}{g} + f^2 ,$$

and

$$\frac{lz^2}{g} = x^2 - f^2 \text{ or, if } l = g,$$

$$z^2 = x^2 + f^2$$

and

$$z^2 = x^2 - f^2 .$$

Which had to be demonstrated here.

322. E L E M. C V R V A R V M

recta D K; ita ut ad easdem diametros ordinatim applicatæ angulos faciant, dato vel assumpto angulo A B E vel D K E sive D X E æquales. Eritque casu utroque D centrum, & semilatus transversum ∞f, quod in dictis diametris respectivè per lineas D V vel D L exprimatur; eritque porrò transversi lateris ad rectum ratio, ut l ad g. Si enim descripta intelligatur prædicta Hyperbola per punctum V in utraque diametro, versùs X & K respectivè, eademque secare supponatur rectam X E, ut & ipsam K E, ad dictas diametros ordinatim applicatas, in puncto E: erit D A X sive K B E $\infty y + c$, & D X seu K E $\infty y - c$; ideoque eadem D A X & K B E vel D X & K E ea ipsa, quæ pro z est assumpta: ac propterea L X $\infty z + f$, & L K $\infty x + f$

atque V X $\infty z - f$, & V K $\infty x - f$:

ideoque rectangula

L X V $\infty zz - ff$, & L K V $\infty xx - ff$.

Cumque eadem sit ratio tam unius quàm alterius rectanguli L X V ad quadratum X E, ut & utriusque rectanguli L K V ad quadratum ex K E vel K B E respectivè, quæ est lateris transversi ad rectum, hoc est, ut l ad g: erit quoque ut

l ad g, ita $zz - ff$ ad xx,

itemque ut l ad g, ita $xx - ff$ ad zz:

hoc est, revocatâ proportione ad æqualitatem,

erit $\frac{lxx}{g} \infty zz - ff$, sive $zz \infty \frac{lxx}{g} + ff$,

& $\frac{lzz}{g} \infty xx - ff$: aut, si l sit ∞g,

erit $zz \infty xx + ff$,

& $zz \infty xx - ff$.

Quod quidem hîc demonstrandum erat.

§. 2.   At verò secundò, si z assumpta sit pro $y \, 8 \, \frac{bx}{a}$, sumpto in linea B E puncto M, ita ut A B ad B M sit, sicut a ad b; hoc est, ut B M sit $\infty \frac{bx}{a}$, (quod quidem punctum M, si z assumpta fuerit pro $y - \frac{bx}{a}$, ab eadem parte lineæ A B quâ datus vel conceptus angulus A B E sumendum est; sed contra, si habeatur $z \infty y + \frac{bx}{a}$, ab altera parte ejusdem lineæ A B sumi debet,) oportet per puncta A & M rectam lineam ducere A M, secantem H C H & Q F Q per prædicta puncta C & F ductas ipsi B E parallelas in punctis G & N.

Quo

[322, cont.]

§2. But in the second case, if z has been adopted instead of $y \pm (bx/a)$ [4.38], we first have to select a point M on BE, so that AB is to BM as a to b, that is so that $BM = bx/a$ (this point M has to be selected on the same side of line AB as where the angle ABE has been given or chosen, if z has been adopted instead of $y - (bx/a)$, but otherwise on the opposite side of this line AB, if we have $z = y + (bx/a)$); next we have to draw the straight line AM through the points A and M, which intersects the lines HCH and QFQ, drawn through the aforementioned points C and F parallel to BE, at the points G and N.

[323]

If the term f^2 is provided with the plus sign, then the diameter of the required hyperbola will be situated on the straight line AW parallel to BE; the ordinate-wise applied lines, such as EW, are parallel to AM.

But otherwise (that is, if the term f^2 is provided with the minus sign) the diameter will be situated on the aforementioned AM so that the ordinate-wise applied lines

L I B. II. C A P. IV.   323

'Quo facto, si terminus *ff* signo **+** affectus sit, erit quæsitæ Hyperbolæ diameter in recta A W ipsi B E parallela, ad quam ordinatim applicatæ, ut E W, sunt ipsi A M æquidistantes. Sin con-

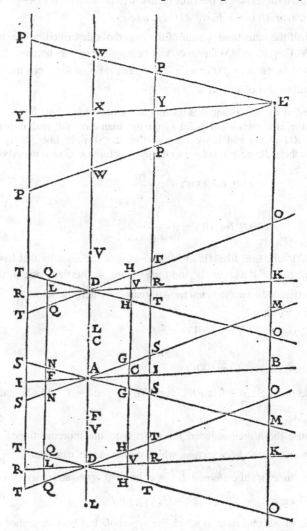

tra, hoc est, si terminus *ff* signo **—** sit affectus, erit diameter in prædicta recta A M, ita ut ordinatim ad eam applicatæ cum ipsa
Ss 2   faciant

[324]

enclose angles with AM that are equal to AME or $AMBE$ with AM. The center of one hyperbola as well as the center of the other will be situated at the point A. And with respect to both their latus transversum and latus rectum: half the latus transversum of the hyperbola described on the diameter AW will be f (this is again expressed by AC or AF) and the ratio of this latus transversum to the latus rectum will be as a^2l is to e^2g. We suppose namely that the ratio of AB to the drawn AM is as a to e.

But half the latus transversum of the hyperbola described on the diameter AM will be AG or AN. This AG or AN, however, will be equal to ef/a, because AC (or AF, i.e. f) is to AG (or AN) as AB is to AM or as a is to e. The ratio of this latus transversum to the latus rectum will be that of e^2l to a^2g.

Indeed, if we understand that the aforementioned hyperbola has been described passing through the aforementioned point C on the diameter AW and through point G on the diameter AM [4.39] and if we suppose that the straight line ME or WE (ordinate-wise applied to these diameters) is intersected by the aforementioned hyperbola at point E, then:

$$MBE \text{ (or } AXW) = y + \frac{bx}{a} \qquad\qquad [4.40]$$

and

$$ME \text{ (or } AW) = y - \frac{bx}{a} \qquad\qquad [4.41];$$

that is, AXW or MBE, like AW or ME, are exactly the quantity that has been adopted as z. However, AM (or WE) $= ex/a$ and furthermore, in the first case, in which the hyperbola has been described on the diameter AW (namely when the term f^2 is provided with the plus sign), we have

$$FW \text{ (or } FXW) = z + f$$

and

$$CW \text{ (or } CXW) = z - f,$$

hence the rectangle satisfies FWC (or $FXWC$) $= z^2 - f^2$ [4.42] and the square on WE equals e^2x^2/a^2.

Because the aforementioned rectangle is to the aforementioned square as the latus transversum is to the latus rectum, which means in this case that $z^2 - f^2$ is to e^2x^2/a^2 as a^2l is to e^2g, therefore $e^2lx^2 = e^2gz^2 - e^2gf^2$ and after division of all terms by e^2g, $lx^2/g = z^2 - f^2$, i.e. $z^2 = (lx^2/g) + f^2$.

But in the second case, in which the hyperbola has been described on the diameter AM (namely when the term f^2 is provided with a minus sign), then

$$NM = \frac{ex}{a} + \frac{ef}{a} \text{ and } GM = \frac{ex}{a} - \frac{ef}{a}$$

faciant angulos angulo A M E vel A M B E æquales : eritque tam
unius quàm alterius Hyperbolæ centrum in puncto A. Et quantùm ad earundem latera tam transversa quàm recta, erit ejus Hyperboles, quæ ad diametrum A W describitur, semi-latus transversum ∞f (idque iterum exprimatur per A C vel A F), & ratio
ejusdem transversi lateris ad rectum, ut aal ad eeg; posito nimirum quòd ratio ipsius A B ad ductam A M sit ut a ad e; at verò
Hyperboles, quæ ad diametrum A M describitur, semi-latus
transversum erit A G vel A N. Quæ quidem A G vel A N erit
$\infty \frac{ef}{a}$; cum sit ut A B ad A M, sive ut a ad e; ita A C vel A F, hoc
est, f, ad A G vel A N; & ratio ejusdem transversi lateris ad rectum, ut eel ad aag. Si enim prædicta Hyperbola descripta intelligatur, transiens per prædictum punctum C in diametro A W
& per punctum G in diametro A M, præsupponaturque rectam
M E vel W E ordinatim ad easdem diametros applicatas à prædicta Hyperbola secari in puncto E : erit M B E vel A X W $\infty y +$
$\frac{bx}{a}$, & M E vel A W $\infty y - \frac{bx}{a}$, hoc est, A X W seu M B E, uti &
A W seu M E ea ipsa erit, quæ pro z assumpta est. Est autem A M
seu W E $\infty \frac{ex}{a}$, ac porrò casu priori, ubi descripta est Hyperbola ad diametrum A W, (cùm nempe terminus ff signo $+$ est affectus) F W sive F X W $\infty z + f$,
   & C W sive C X W $\infty z - f$:
      ideoque rectangulum

   F W C vel F X W C $\infty zz - ff$, & quadratum W E $\infty \frac{eexx}{aa}$.

   Cumque sit ut latus transversum ad rectum, ita prædictum rectangulum ad prædictum quadratum, hoc est, eo casu ut aal ad
eeg, ita $zz - ff$ ad $\frac{eexx}{aa}$: erit $eelxx \infty eegzz - eegff$, &, omnibus per eeg divisis, $\frac{lxx}{g} \infty zz - ff$, id est, $zz \infty \frac{lxx}{g} + ff$.

   At verò casu posteriori, ubi descripta est Hyperbola ad diametrum A M, cùm nempe terminus ff signo $-$ est affectus, erit
N M $\infty \frac{ex}{a} + \frac{ef}{a}$, & G M $\infty \frac{ex}{a} - \frac{ef}{a}$: ideoque rectangulum N M G
$\infty \frac{eexx}{aa} - \frac{eeff}{aa}$. Cumque sit ut latus transversum ad rectum, id
est, hoc casu, ut eel ad aag, ita prædictum rectangulum N M G
                                                                      ad

[324, cont.]

and therefore the rectangle *NMG* satisfies

$$NMG = \frac{e^2 x^2}{a^2} - \frac{e^2 f^2}{a^2}.$$

Because the aforementioned rectangle *NMG* is to the square on *ME* or *MBE* (that is to z^2),

[325]

as the latus transversum is to the latus rectum, that is, in this case as $e^2 l$ to $a^2 g$, therefore $(e^2 x^2 - e^2 f^2)/a^2$ is to z^2 as $e^2 l$ is to $a^2 g$ and, therefore,

$$e^2 l z^2 = e^2 g x^2 - e^2 g f^2.$$

This means that after division by $e^2 g$,

$$\frac{l z^2}{g} = x^2 - f^2.$$

Which was here to be demonstrated.

ad M E vel M B E quadratum, hoc est, ad zz : erit ut ee l ad aag,
ita $\frac{cexx - ecff}{aa}$ ad zz : ac proinde $eelzz \infty eegxx - eegff$.

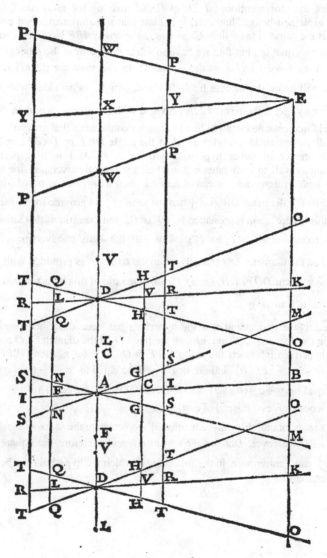

Hoc est, factâ divisione per eeg, erit $\frac{lzz}{g} \infty xx - ff$. Quod hîc
demonstrandum erat.

Si.

[326]

§3. If finally, in the third case, z has been adopted instead of $y \pm (bx/a) \pm c$, then, as before, we have first to draw AD equal to c (the text gives f instead of c, transl.) [4.43] and DK parallel to AB and to select point O on line KE so that DK is to KO as a is to b, i.e. so that KO is bx/a. Next we have to draw the straight line DO through the points D and O intersecting the aforementioned HCH at H and meeting the aforesaid QFQ at Q. (From what we explained before, however, it is clear that the aforementioned point O has to be selected at the same side of line AB as where the angle ABE has been chosen if we have $-bx/a$ in the equation, but that we have to select this point at the other side of this line if we have if we have $+bx/a$ in the equation). In this way the diameter of the required hyperbola will be on the straight line DW if the term f^2 is provided with the plus sign. If on the contrary, i.e. if the term f^2 is provided with the minus sign, then the diameter will be on the aforementioned straight line DO, such that the lines that are ordinate-wise applied to these diameters enclose angles equal to the angle DWE or $DXWE$, or DOE or $DOKE$ [4.44]; the center of either hyperbola will be at D. And with respect to their latus transversum as well to their latus rectum: half the latus transversum (the text erroneously gives the *latus transversum* instead of *half the latus transversum*, transl.) of the hyperbola described on the diameter DW (i.e. when the term f^2 is provided with the plus sign) will be f and this is here again represented by DV or DL and the ratio of this latus transversum to the latus rectum is as $a^2 l$ to $e^2 g$. But half the latus transversum of the hyperbola described on the diameter DO (namely when the term f^2 is provided with the minus sign) will be the segment DQ or DH, i.e. ef/a and the ratio of this latus transversum to the latus rectum is as $e^2 l$ to $a^2 g$.

Indeed, if we understand that the hyperbola has been described passing through the point V on the diameter DW and through the point H on the diameter DO and if we suppose that this hyperbola intersects the straight WE or OE at point E, then $OKBE$ or $DAXW$ will be equal to $y + c + (bx/a)$ and OE or DW will be equal to $y - c - (bx/a)$; OBE or DAW will be equal to $y + c - (bx/a)$ and OKE or DXW will be equal to $y - c + (bx/a)$ (the text erroneously gives d instead of c, transl.).

This means that all these aforementioned segments are the same as those that have been adopted as z. However, DO (or WE) $= ex/a$ and therefore the square on WE equals $e^2 x^2/a^2$ and, furthermore, in the first case (in which the hyperbola has been described on the diameter DW,

§. 3.   Si denique tertiò z assumpta sit pro $\gamma \, \beta \, \frac{bx}{a} \, \beta \, c$, ductâ, ut supra, A D ∞ f, & D K ipsi A B parallelâ, sumptoque in linea K E puncto O; ita ut D K ad K O sit, sicut a ad b, hoc est, ut K O sit $\infty \, \frac{bx}{a}$, ducenda est per puncta D & O recta D O, secans prædictam H C H in H, atque occurrens præfatæ Q F Q in Q. (Constat autem ex superiùs explicatis prædictum punctum O, si in æquatione habeatur $- \frac{bx}{a}$, ab eadem parte lineæ A B sumendum esse, quâ datus aut assumptus est angulus A B E; at si habeatur $+ \frac{bx}{a}$, illud ipsum punctum ex altera ejusdem lineæ parte sumi debere.) Quo facto, si terminus ff signo $+$ affectus sit, erit diameter quæsitæ Hyperbolæ in recta D W. Sin contra, hoc est, si terminus ff signo $-$ sit affectus, erit ipsa in prædicta recta D O; ita ut ad easdem diametros ordinatim applicatæ angulos faciant angulo D W E sive D X W E, aut D O E sive D O K E æquales: eritque tam unius quàm alterius Hyperbolæ centrum in puncto D. Et quantùm ad earundem latera tam transversa quàm recta, erit ejus Hyperbolæ, quæ ad diametrum D W describitur, hoc est, cùm terminus ff signo $+$ afficitur, latus transversum ∞ f, idque hîc iterum exprimatur per D V vel D L, ac ratio ejusdem lateris transversi ad rectum, ut $a a l$ ad $e e g$; at verò Hyperboles, quæ ad diametrum D O describitur, nimirum, quando terminus ff signo $-$ affectus est, erit semi-latus transversum recta D Q vel D H, id est, $\frac{ef}{a}$; atque ratio ejusdem lateris transversi ad rectum, ut $e e l$ ad $a a g$. Si enim descripta intelligatur Hyperbola, transiens per punctum V in diametro D W & per punctum H in diametro D O, supponaturque eandem Hyperbolam secare rectam W E vel O E in puncto E, erit O K B E sive D A X W ∞ $y + c +$ $\frac{bx}{a}$, & O E sive D W ∞ $y - c - \frac{bx}{a}$, ac O B E sive D A W ∞ $y + c$ $- \frac{bx}{a}$, atque O K E vel D X W ∞ $y - d + \frac{bx}{a}$. Hoc est, erunt omnes illæ prænominatæ lineæ eædem, quæ pro z assumptæ sunt. Est autem D O seu W E ∞ $\frac{ex}{a}$, ideoque quadratum W E ∞ $\frac{e e x x}{a a}$: ac porrò casu priori, ubi descripta est Hyperbola ad diametrum D W,

[327]

when namely the term f^2 is provided with the plus sign) LW (or LXW) $= z + f$ and VW (or VWX) $= z - f$. Therefore, the rectangle LWV (or $LXWV$) equals $z^2 - f^2$.

Now the aforementioned rectangle is to the square on WE as the latus transversum to the latus rectum, which means in this case that $z^2 - f^2$ is to $e^2 x^2 / a^2$ as $a^2 l$ is to $e^2 g$.

L I B. II. C A P. IV.   327

D W, cùm nempe terminus *ff* signo + afficitur, L W sive
L X W ∞ z + *f*, & V W sive V X W ∞ z — *f* : ideoque rectangulum L W V sive L X W V ∞ z z — *ff*. Cumque sit ut latus

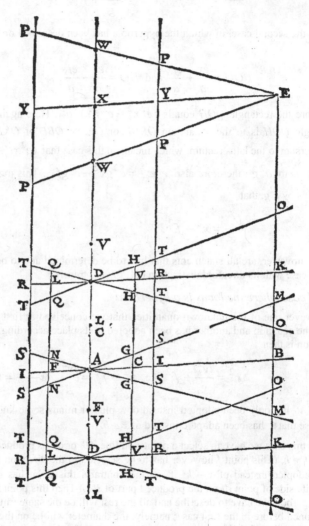

transversum ad rectum, ita prædictum rectangulum ad W E
quadratum, hoc est, eo casu, ut *a a l* ad *e e g*, ita z z — *ff* ad
e e x x

[328]

Therefore: $e^2 lx^2 = e^2 gz^2 - e^2 gf^2$

and, after division of all terms by $e^2 g$,

$$\frac{lx^2}{g} = z^2 - f^2 \text{ or } z^2 = \frac{lx^2}{g} + f^2 .$$

But in the second case, in which the hyperbola has been described on the diameter DO, we have:

$$QO = \frac{ex}{a} + \frac{ef}{a} \text{ and } HO = \frac{ex}{a} - \frac{ef}{a},$$

therefore the rectangle QOH equals $(e^2 x^2 - e^2 f^2)/a^2$. Here again the aforementioned rectangle QOH is to the square on $OKBE$ or OE, or OBE or OKE [4.45] as the latus transversum to the latus rectum, which means in this case that $(e^2 x^2 - e^2 f^2)/a^2$ is to z^2 as $e^2 l$ is to $a^2 g$; therefore also $e^2 lz^2 = e^2 gx^2 - e^2 gf^2$. This means, after division of all terms by $e^2 g$, that

$$\frac{lz^2}{g} = x^2 - f^2 .$$

These, however, are all statements that had to be determined and to be demonstrated in the above case with its threefold distinction (§1, §2, §3, transl).

Third case where the locus is a hyperbola

If however one of the unknown quantities that we conceived at first has been taken away from the equation and another has been adopted in its place according to the Rule and if the equation is then

$$y^2 = \frac{lv^2}{g} + f^2 , \text{ i.e. } y^2 - f^2 = \frac{lv^2}{g} , \text{ or } \frac{ly^2}{g} = v^2 - f^2$$

and if v has only been adopted instead of x plus or minus some known quantity then we suppose that v has been adopted instead of $x \pm h$.

In this case we have to select a point I on line AB, or on its produced part, in such a way that $AI = h$. (This point I however has to be selected in the direction from A to B if v has been adopted instead of $x - h$; if, on the contrary, this point has to be selected on the opposite side of point A, on the produced part of BA.) Then this point I will be the center of the hyperbola we have to describe and all the rest will be the same (mutatis mutandis) as we mentioned before in the first case: namely, the diameter will be on the straight line IY or on the straight line IB, half the latus transversum equals f and the ratio of the latus transversum to the latus rectum will be as l is to g.

328   E L E M.   C V R V A R V M

$\frac{eexx}{aa}$: erit $eelxx \infty eegzz - eegff$, ac, divisis omnibus per eeg,

$\frac{lxx}{g} \infty zz - ff$, sive $zz \infty \frac{lxx}{g} + ff$.

At verò casu posteriori, ubi descripta est Hyperbola ad diametrum D O, erit $QO \infty \frac{ex}{a} + \frac{ef}{a}$, & $HO \infty \frac{ex}{a} - \frac{ef}{a}$; ideoque rectangulum $QOH \infty \frac{eexx - eeff}{aa}$. Cumque iterum sit, ut latus transversum ad rectum, ita prædictum rectangulum Q O H ad quadratum ex O K B E vel O E, sive O B E aut O K E: id est, eo casu, ut $eelad aag$, ita $\frac{eexx - eeff}{aa}$ ad zz: erit quoque proinde $eelzz \infty eegxx - eegff$. Hoc est, divisis omnibus per eeg, erit $\frac{lzz}{g} \infty xx - ff$. Quæ quidem omnia sunt, quæ casu superiori in triplici sua distinctione determinanda ac demonstranda erant.

Casus 3tius, cùm Locus est Hyperbola.   Si verò quantitatum incognitarum ab initio conceptarum, alterâ ex æquatione sublatâ, aliâque ejusdem loco secundùm Regulam assumptâ, æquatio sit $yy \infty \frac{lvv}{g} + ff$, (id est, $yy - ff \infty \frac{lvv}{g}$) aut $\frac{lyy}{g} \infty vv - ff$; atque ipsa v tantùm assumpta sit pro $x \, \beta$ notâ aliquâ quantitate, Sit v assumpta pro $x \, \beta \, h$; Hoc casu in linea A B vel eadem productâ sumendum est punctum I, ita ut A I sit ∞h (quod quidem punctum I, si v assumpta fuerit pro $x - h$, ab A versùs B; Sin contra, ab altera parte puncti A in producta B A sumi debet.) Quo facto, erit idem illud punctum I centrum describendæ Hyperboles, &, mutatis mutandis, cætera omnia, ut supra casu 1mo memoratum est, nempe, diameter in recta I Y vel in recta I B, semi-latus transversum ∞f, atque proportio lateris transversi ad rectum, ut l ad g.

Casus 4ius, cùm Locus est Hyperbola.   Si denique quantitatum incognitarum, primò conceptarum, utrâque ex æquatione sublatâ, aliisque earundem loco juxta Regulam assumptis, æquatio sit $zz \infty \frac{lvv}{g} + ff$, (id est, $zz - ff \infty \frac{lvv}{g}$), aut $\frac{lzz}{g} \infty vv - ff$; atque z primùm assumpta sit pro $y \, \beta \, c$, ducenda est utrinque I R parallela B E, & ∞c: quo facto, erit idem illud punctum R centrum, & diameter in recta R Y vel

[328, cont.]

The fourth case where the locus is a hyperbola

If, finally, either unknown quantity we conceived at first has been taken away from the equation and others have been adopted in their place according to the Rule and if then the equation is

$$z^2 = \frac{lv^2}{g} + f^2, \text{ i.e. } z^2 - f^2 = \frac{lv^2}{g}, \text{ or } \frac{lz^2}{g} = v^2 - f^2.$$

§1. And if z has at first been adopted instead of $y \pm c$ [4.46], then we have to draw IR parallel to BE and equal to c, on either side (of $DAWX$ and AB, transl.). Then this point R will be the center and the diameter will be on the straight line RY or RK;

[329]

half its latus transversum will be equal to f and the ratio of the latus transversum to the latus rectum will be as l is to g, just as all these things have been explained at greater length in the second case §1 (mutatis mutandis) [4.47].

§2. But if z has been adopted instead of $y \pm (bx / a)$, then the center (of the conic section) will be at S: that is, the point

L I B. II. C A P. IV.   329

vel R K, ejúfque femi-latus tranfverfum ∞f, ac ratio tranfverfi lateris ad rectum, ut *l* ad *g*. quemadmodum ea omnia, mutatis mutandis, cafu fecundo §. 1. fufiùs explicata funt.

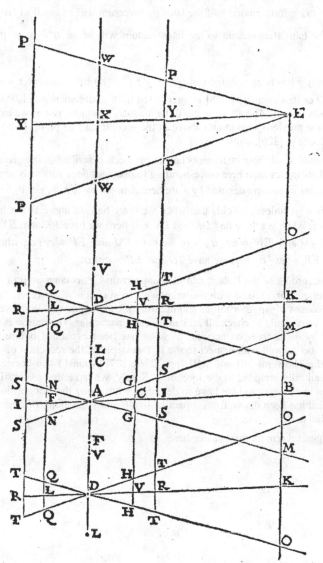

§. 2.   At fi z affumpta fuerit pro *y* ℞ $\frac{bx}{a}$, erit punctum S, in

Pars II.   T t   quo

[330]

where MA or its conjugate [4.48] is intersected by the aforementioned IR or, if necessary, by its produced part, and all other things will be the same (mutatis mutandis) as we mentioned before in the second case, §2 (p. [322], transl.). Namely, the diameter of the section will be on the straight line SP or SM (and whereas in the former case AM or EW was equal to ex/a, here SM or EP will be ev/a, because BI, i.e. v, is to SM as AB is to AM, i.e. as a is to e); furthermore, half the latus transversum will be equal to f or ef/a and the ratio of the latus transversum to the latus rectum will be as a^2l to e^2g or as e^2l to a^2g.

§3. If finally z has been adopted instead of $y \pm (bx/a) \pm c$, the center will be at point T, where DO or its conjugate [4.48] is intersected by the aforementioned IR or, if necessary, by its produced part. And all the rest will be, mutatis mutandis, just as we explained at more length in the previous section and above in the second case §3 [4.48] (where the locus is a hyperbola; see p. [326], transl.).

The proof of all these statements has amply been described in the preceding, while all terms and quantities used here corresponding to previous ones with only one exception: the quantities that were there denoted by x are here denoted by $x \pm h$, i.e. v.

In this way, indeed, it holds that when we there had AB and $EX = x$, here we have IB and $EY = v$; when we there had DK and $EX = x$, here we have RK and $EY = v$; when we there had AM and $EW = ex/a$, here we have SM and $EP = ev/a$; when we there had DO and $EW = ex/a$, here we have TO and $EP = ev/a$.

But, according to the Rule it can also happen that v is composed of x plus or minus some other quantity, mixed with the unknown y in such a way, however, that in this case z can only consist of y plus or minus another totally known quantity. We nevertheless think it not at all worthwhile to check all related cases in particular. The reason is that these cases are clear by themselves on account of what has been explained before with respect to parabolic loci as well with respect to the last example of the reduction of equations to the form of the Theorems XII and XIII (pp. [276], [277], transl.) and because they are also entirely and fully included in the foregoing. Surely, we have first to substitute x for y and vice versa in all cases and we have to delineate x not along the straight line AB, but along the line that has been drawn from A parallel to BE and y not along BE, but along a straight line parallel to AB.

This general remark may suffice here.

330   E L E M.   C V R V A R V M

quo M A , vel quæ ipſi in directum adjungitur, per prædictam
I R, vel eandem productam, ſi opùs ſit, interſecatur, centrum ſectionis; & cætera omnia, mutatis mutandis, ut ſupra caſu ſecundo
§. 2. memoratum eſt. Nempe erit ſectionis diameter in recta S P
vel S M(atque ut ibidem A M ſeu E W erat $\infty \frac{ex}{a}$, ita hîc S M ſeu

E P erit $\infty \frac{ev}{a}$: cum ſit ut A B ad A M, hoc eſt, ut a ad e, ita B I,
hoc eſt, v, ad S M); eritque porrò ſemi-latus transverſum ∞f &
$\frac{ef}{a}$ reſpective, ac ratio transverſi lateris ad rectum, ut $a\,a\,l$ ad $e\,e\,g$,
vel ut $e\,e\,l$ ad $a\,a\,g$.

§. 3.   Si denique z aſſumpta fuerit pro y ꝗ $\frac{bx}{a}$ ꝗ c, erit punctum T,
in quo D O, vel quæ ipſi in directum adjungitur, per prædi-
ctam I R, vel productam, ſi opùs ſit, interſecatur, centrum; &
reliqua omnia, mutatis mutandis, ut paragrapho præcedenti,
& ſupra caſu ſecundo §. 3. fuſiùs expoſitum eſt. Atque eorum
omnium demonſtratio in præcedentibus explicitè eſt compre-
henſa, cum termini & quantitates omnes hîc cum prioribus conveniant, excepto tantùm, quòd, quæ ibidem deſignabantur per x,
hîc ſint x ꝗ b, hoc eſt, v. Ita enim quod ibi erat A B & E X ∞ x,
hîc eſt I B & E Y ∞ v; quod ibi erat D K & E X ∞ x, hîc eſt R K
& E Y ∞ v; quod ibi erat A M & E W $\infty \frac{ex}{a}$, hîc eſt S M & E P ∞
$\frac{ev}{a}$; quod ibi erat D O & E W $\infty \frac{ex}{a}$, hîc eſt T O & E P $\infty \frac{ev}{a}$.

Quamvis autem ſecundùm Regulam accidere etiam poſſit, ut
v compoſita ſit ex x ꝗ aliâ quâdam quantitate, cui & incognita y
permixta ſit; ita tamen, ut eo caſu z ſolummodo ex y ꝗ aliâ quantitate in totum cognitâ conſtare queat, haudquaquam tamen operæ
pretium exiſtimamus, caſus omnes eò ſpectantes ſpeciatim perſe-
qui: cum ex iis, quæ tam in Locis Parabolicis quàm in poſteriori
exemplo reductionis æquationum ad formulas Theorematum
12^{mi} & 13^{tii} ſuperiùs explicata ſunt, iidem illi caſus per ſe manifeſti
ſint atque in præcedentibus etiam omnino plenèque comprehen-
dantur, ſi nimirum, ſubſtituto per omnia x loco y & vice verſâ, eadem x non per rectam A B ſed per eam, quæ ex A ipſi B E parallela ducta ſit, atque y non per B E ſed per rectam ipſi A B æqui-
diſtantem, deſignetur. Quòd hîc generaliter monuiſſe ſuffecerit.

Alii

[331]

Four other cases where the locus is a hyperbola

Moreover, we remarked before that it can also happen that the equation is

1. $yx = f^2$,

2. $zx = f^2$,

3. $yv = f^2$,

4. $zv = f^2$,

and that in all these cases the required locus is a hyperbola, whose determination or description follow straightforward from what has been explained before.

Indeed, if in the first case we measure the segment AC equal to f, along the straight line AB, and if we then raise the segment CD from the point C parallel to BE and equal to the aforementioned segment AC, i.e. f, and if we finally draw a straight line through A and D, then A will be the center of a hyperbola whose axis is on the straight line AD and whose vertex is point D and whose asymptote is AB. In other words: if we first draw the segment DF perpendicular to AD and ending on AB, then AD will be half the latus transversum and the ratio of the latus transversum to the latus rectum will be as the square on AD to the square on DF.

For if we suppose that the aforementioned hyperbola intersects the straight line BE at point E, then[1] the rectangle ABE will be equal to the square on AC or on CD [4.49]. Because $AB = x, BE = y$, and $AC = f$, therefore $xy = f^2$. Which was to be demonstrated in the first case.

In he second case, in which as we know, the equation is $zx = f^2$, z has to be adopted, according to the Rule, instead of y plus or minus some known quantity.

---

[1] Prop. 3, Lib. I, p. [180].

Lib. II. Cap. IV. 331

Alii quatuor casus, cùm Locus est Hyperbola.

Iam verò quod supra annotavimus accidere quoque posse, ut æquatio sit

1. $yx \infty ff$,
2. $zx \infty ff$,
3. $yv \infty ff$,
4. $zv \infty ff$,

omnibusque istis casibus Locum quæsitum esse Hyperbolam, ejus determinatio sive descriptio atque demonstratio ex iis, quæ jam ante explicata sunt, sponte quoque profluunt.

Primo enim casu, si in recta A B sumatur A C ∞f, atque ex puncto C eductâ rectâ C D, quæ ipsi B E sit æquidistans & æqualis priori A C, hoc est ∞f, per A & D recta linea ducatur: erit A centrum Hyperbolæ, cujus axis est in recta A D, & punctum D vertex, atque A B asymptotos: sive (ductâ rectâ D F ad A D perpendiculari ac in A B terminata) erit A D semi-latus transversum, & ratio transversi ad rectum, ut A D quadratum ad D F

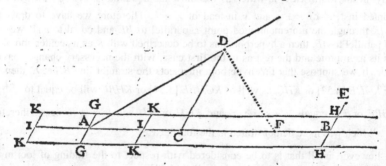

quadratum. Si namque prædicta Hyperbole secare supponatur rectam B E in puncto E, erit [1] rectangulum A B E ∞ quadrato ex A C vel C D. Quare cum A B sit ∞x, B E ∞y, & A C ∞f: erit $xy \infty ff$. Quod primo casu erat demonstrandum. _{[1] *per 3 primi hujus.*}

Secundo casu, cùm nempe æquatio est $zx \infty ff$, oportet ut z juxta Regulam sit assumpta pro y ꝑ notâ quâdam quantitate. Esto

T t 2 itaque

[332]

Now let z have been adopted instead of $y \pm c$. To describe the hyperbola we then have to draw the segment AG through point A, parallel to BE and equal to c. This point G is, of course, selected either on this side of the line AB, or on the other depending on whether the quantity c is provided with the plus sign or the minus sign. If, furthermore, we draw GH parallel to AB, then the hyperbola has to be described with G as its center and GH as one of its asymptotes, the rest (mutatis mutandis) being the same as before.

If we now suppose that this curve intersects BE at point E, then the rectangle $GHBE$ [4.50] or GHE will be equal to f^2. As $GH = x$ and HE (or HBE) $= y \mp c$, i.e. z (the text erroneously gives $\pm$, transl.), then the rectangle GHE or $GHBE$ will be equal to zx and so $zx = f^2$. Which was to be demonstrated in the second case.

In the third case, in which as we know, the equation is $yv = f^2$, v has only to be adopted instead of x plus or minus some known quantity, say instead of $x \pm h$.

Therefore, to find the required locus, we have to measure the segment AI, equal to h, along the straight line AB or along its produced part and hereafter a hyperbola is to be described with I as its center and IAB or IB as its asymptote [4.51], and the rest as before, with the necessary changes having been made. If we suppose that this hyperbola intersects the straight line BE at E, then the rectangle $IABE$ or IBE will be equal to f^2 [4.52]. As IAB (or IB) equals $x \pm h$, i.e. v, and as $BE = y$, therefore $yv = f^2$.

Which was to be demonstrated in the third case.

Finally, in the fourth case, in which as we know the equation is $zv = f^2$, z will have been adopted instead of $y \pm c$ and v instead of $x \pm h$. Therefore we have to draw the segment IK through the aforementioned point I, parallel to BE and equal to c. If we first draw KH parallel to AB, then a hyperbola has to be described with K as its center and KGH or KH as its asymptote and the rest as in the first case, with the necessary changes having been made. If we suppose that this hyperbola intersects the straight line BE at E, then the rectangle $KGHE$ [4.53] or KHE, as well as $KGHBE$ [4.54] or $KHBE$ will be equal to f^2.

As HBE or HE equals $y \pm c$, i.e. z, and KGH or KH equals $x \pm h$, i.e. v, therefore $zv = f^2$. Which was to be demonstrated in the fourth case

This, however, is all that is to be considered with respect to the finding of loci in the case in which they are situated on a hyperbolic line [4.54].

In the other general case, however, of the formulas included in number 3 (p. [305], transl.) [4.55], in which as we know the term x^2 or v^2 is found provided with the minus sign and, hence, the required locus appears to be an ellipse or the circumference of a circle, a fraction may be found in the equation. In this case this fraction can also be transferred to the term y^2 or z^2 for clarity's sake.

332 ELEM. CVRVARVM

itaque affumpta pro $y \propto c$, atque idcirco ad defcribendam Hyperbolam ducatur per punctum A recta A G ipfi B E parallela, ac ∞ c: fumpto nimirum puncto G vel ab hac vel ab illa parte lineæ A B, prout c quantitas figno + vel — fuerit affecta; ductâque porrò G H ipfi A B parallelâ, centro G, Afymptoto G H, cæterisque, ut fupra, mutatis mutandis, Hyperbole defcribatur. Hæc igitur fi fecare fupponatur rectam B E in puncto E, erit rectangulum G H B E vel G H E $\propto ff$. Vnde cum fit G H ∞ x, & H E vel H B E $\propto y \propto c$, id eft, z : erit G H E vel G H B E rectangulum ∞ zx, ac propterea $zx \propto ff$. Quod 2do cafu demonftrandum erat.

Tertio cafu, nempe fi æquatio fit $y v \propto ff$: v quoque tantùm pro $x \propto$ notâ quâdam quantitate fumpta fit oportet, veluti pro $x \propto h$. Ideoque ad inventionem Loci quæfiti, in recta A B vel in ipfâ productâ fumenda eft A I ∞ h, ac porrò centro I, atque Afymptoto I A B vel I B, cæterisque, ut fupra, mutatis mutandis, defcribenda eft Hyperbola, quæ fi rectam B E fecare fupponatur in E: erit rectangulum I A B E vel I B E $\propto ff$. Quare cum I A B vel I B fit ∞ $x \propto h$, hoc eft, v, & B E ∞ y: erit $y v \propto ff$. Quod 3^{tio} cafu demonftrandum erat.

Denique quarto cafu, fi nempe æquatio fit $z v \propto ff$: erit z affumpta pro $y \propto c$, & v pro $x \propto h$. Ideoque per prædictum punctum I ducenda eft I K ipfi B E æquidiftans & ∞ c; ductâque K H ipfi A B parallelâ, centro K, atque Afymptoto K G H vel K H, cæterisque, ut cafu 1mo, mutatis mutandis Hyperbole defcribenda eft, quæ fi fecare fupponatur rectam B E in E: erit rectangulum K G H E vel K H E, ut & K G H B E vel K H B E $\propto ff$. Hinc cum H B E vel H E fit ∞ $y \propto c$, id eft, z, & K G H vel K H ∞ $x \propto h$, hoc eft, v: erit $z v \propto ff$. Quod 4to cafu demonftrandum erat.

Atque hæc quidem omnia funt, quæ circa inventionem Locorum eo cafu, quo iidem funt in linea Hyperbolica, confideranda veniunt.

Altero autem cafu generali formularum fub N^{ro} 3. comprehenfarum, cùm nempe terminus, in quo invenitur xx vel vv figno — fit affectus, ac proinde Locus quæfitus vel Ellipfis vel Circuli circumferentia exiftit, fi in æquatione fractio reperiatur, rejici quoque illa poterit majoris perfpicuitatis gratiâ in terminum yy vel zz. Quo facto primò, remanente utrâque quantitate
inco-

[333]

First case where the locus turns out to be either an ellipse or the circumference of a circle

After these preparations and if we maintain either unknown quantity that we conceived from the beginning, the equation will present itself in the following form:

$$\frac{ly^2}{g} = f^2 - x^2.$$

And, as the following figure shows, the diameter of the ellipse to be described will be situated on the straight line *AB*, that has indeterminately been conceived as *x*, in such a way that the lines that are ordinate-wise applied to this diameter include angles equal to the given or chosen angle *ABE*; its center is at point *A* and half the latus transversum equals *f*, which is expressed by the segment *AC* or *AF* on the aforementioned diameter and the ratio of this latus transversum to the latus rectum will be as *l* to *g*.

Indeed, if we assume that the aforementioned ellipse has been described passing through the points *C* and *F* and intersecting the applied *BE* at point *E*, then $FB = f + x$ and $BC = f - x$, and therefore the rectangle *FBC* equals $f^2 - x^2$.

By the nature of an ellipse, the aforementioned rectangle *FBC* [1] equals the square on *BE*, i.e. y^2, if the latus rectum and the latus transversum are equal [4.56]; therefore in this case $y^2 = f^2 - x^2$ also holds and it is easy to see that the aforementioned curve will be the circumference of a circle if, moreover, with the same suppositions *BE* would also be perpendicular to the segment *FC*, which means that the angle enclosed by the ordinate-wise applied lines and the diameter is a right angle [3.5].

Furthermore, the following holds: if the latus transversum and the latus rectum are

---

[1] Prop. 13, Lib. I, p. [205].

L I B. II. C A P. IV. 333

incognitâ ab initio conceptâ, ſequenti formulâ ſe exhibebit æ-
quatio $\frac{lyy}{g} \infty ff - xx$: eritque,ut in ſequenti figura, deſcribendæ *Caſus*
1^{mus}, cùm
Ellipſeos diameter in recta A B,quæ pro *x* indeterminatè eſt con- *Locus vel*
cepta, ita ut ad eandem diametrum ordinatim applicatæ cum ea *Ellipſis*
vel Circu-
angulos faciant, dato vel aſſumpto angulo A B E æquales; ac cen- *li circum-*
trum in puncto A, & ſemi-latus tranſverſum ∞f. id quod in di- *ferentia*
cta diametro per lineam A C vel A F exprimatur, eritque ratio *exiſtit.*
ejuſdem tranſverſi lateris ad rectum, ut *l* ad *g*.

Si enim deſcripta intelligatur prædicta Ellipſis, tranſiens per
puncta C & F, ſecanſque applicatam B E in puncto E: erit F B
$\infty f + x$, & B C $\infty f - x$: ideoque rectangulum F B C $\infty ff - x x$.
At cum ex natura Ellipſeos, lateribus recto tranſverſoque æqualibus, prædictum rectangulum F B C [1] ſit ∞ quadrato ex B E, [1] *per 13*
hoc eſt, yy: erit quoque proinde eo caſu $yy \infty ff - xx$. Et facilè *primi hu-*
jus.

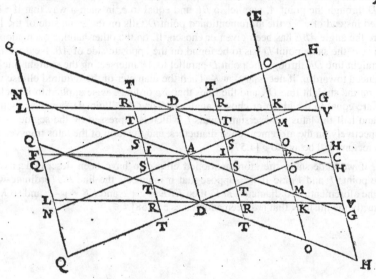

apparet, ſi, iiſdem poſitis, B E ſuper rectam F C foret quoque
perpendicularis, hoc eſt, ut angulus quem ordinatim applicatæ
faciunt ad diametrum ſit rectus, prædictam curvam fore Circuli
circumferentiam.

Cum autem porrò, lateribus tranſverſo rectoque inæqualibus

[334]

unequal and if their ratio is as l to g, then the ratio[1] of the rectangle FBC to the square on BE is the same as the ratio of the latus transversum to the latus rectum, i.e. as l to g. Therefore it is clear from the preceding that $f^2 - x^2$ will be to y^2 as l is to g, which means that $ly^2 / g = f^2 - x^2$. Which was to be demonstrated in this case.

Second case where the locus is either an ellipse or the circumference of a circle

But if one of the unknown quantities that we conceived at first, has been taken away from the equation and if another has been adopted in its place according to the Rule, and if then the equation is

$$\frac{lz^2}{g} = f^2 - x^2,$$

then z will have been adopted instead of $y \pm c$, or instead of $y \pm (bx / a)$ or instead of $y \pm c \pm (bx / a)$.

§1. Indeed, in the first case, if z has been adopted instead of $y \pm c$, we have to draw the segment AD through the point A, parallel to BE and equal to c, in such a way that if z has been adopted instead of $y - c$ the aforementioned point D falls on the same side of the line AB as where the angle ABE has been given or chosen. If, on the other hand, z been adopted instead of $y + c$, the same point D has to be found on the opposite side of AB. Next we first draw the straight line DK through the point D parallel to AB intersecting the straight line BE (being produced toward B, if necessary) at K. Then the diameter of the required ellipse will be situated on the straight line DK and the lines that are ordinate-wise applied to it enclose angles that are equal to the given or chosen angle ABE or DKE. Point D, however, will be the center and half the latus transversum equals f, which is expressed by the segments DV and DL respectively on the aforementioned diameters, and the ratio of the latus transversum to the latus rectum will be as l to g [4.57].

Indeed, if we understand the aforementioned ellipse to have been described passing through the points L and V and if we suppose that it intersects the line BE (ordinate-wise applied to the aforementioned diameter) at E, then $KBE = y + c$ and $KE = y - c$ and so KBE or KE is exactly the quantity that has been adopted as z [4.58].

---

[1] Prop. 13, Lib. I, p. [205].

_{1 per 13}
_{primi hu-}
_{jus.}

atque in ratione ut *l* ad *g*, eadem sit ratio ¹ rectanguli F B C ad B E quadratum, quæ est lateris transversi ad rectum, hoc est, ut *l* ad *g*: ex prædictis palàm est fore ut *l* ad *g*, ita *ff — x x* ad *yy*, hoc est, esse $\frac{lyy}{g}$ ∞ *ff — xx*. Quod eo casu demonstrandum erat.

Casus
2ᵘˢ, cùm
Locus est
vel Ellipsis
vel Circuli
circumfe-
rentia.

At si, quantitatum incognitarum primò conceptarum unâ ex æquatione sublatâ aliâque in ejusdem locum juxta Regulam assumptâ, æquatio sit $\frac{lzz}{g}$ ∞ *ff — xx*: aut *z* assumpta erit pro *y* ප *c*, aut pro ප $\frac{bx}{a}$, aut pro *y* ප *c* ප $\frac{bx}{a}$.

§. 1.   Et primùm quidem, si *z* assumpta fuerit pro *y* ප *c*, ducenda est per punctum A recta A D ipsi B E parallela ac ∞ *c*, ita ut, si *z* fuerit assumpta pro *y — c*, prædictum punctum D cadat ab eadem parte lineæ A B, quâ datus vel conceptus est angulus A B E; sin contra *z* fuerit assumpta pro *y + c*, idem illud punctum D ab altera parte lineæ A B reperiatur. Deinde ductâ per D rectâ D K ipsi A B parallelâ, quæ secet rectam B E, productam versùs B, si opùs fuerit, in puncto K, erit quæsitæ Ellipseos diameter in recta D K, ad quam ordinatim applicatæ cum ea angulos faciant, dato vel assumpto angulo A B E seu D K E æquales. Punctum autem D centrum erit, & semi-latus transversum ∞ *f*. quod in dictis diametris per lineas D V & D L exprimatur, eritque ratio transversi lateris ad rectum, ut *l* ad *g*.

Si enim prædicta Ellipsis descripta intelligatur transiens per puncta L & V, quæ supponatur secare rectam B E, ad prædictam diametrum ordinatim applicatam, in puncto E: erit K B E ∞ *y + c*, & K E ∞ *y — c*, ideoque eadem K B E vel K E ea ipsa, quæ pto *z* assumpta est. Cumque L K sit ∞ *f + x*, & K V ∞ *f — x*: erit rectangulum L K V ∞ *ff — xx*. At cum eadem sit ratio dicti rectanguli L K V ad quadratum ex K B E vel K E, hoc est, ad *z z*, quæ est lateris transversi ad rectum, hoc est, ut *l* ad *g*: erit ut *l* ad *g*, ita *ff — x x* ad *z z*, hoc est, erit $\frac{lzz}{g}$ ∞ *ff — x x*. Quod quidem, si *l* sit ∞ *g*, idem est ac *z z* ∞ *ff — xx*. Atque hîc iterum facilè apparet, quòd, existente angulo D K B E vel D K E recto, & *l* ∞ *g*, hoc est, rectangulo L K V ∞ K E quadrato, prædicta curva Circulus sit futura.

§. 2.   At verò, si *z* assumpta fuerit pro *y* ප $\frac{bx}{a}$, sumpto in linea B E,

pro-

[334, cont.]

As $LK = f + x$ and $KV = f - x$, the rectangle LKV will therefore be equal to $f^2 - x^2$.

But the ratio of the aforementioned rectangle LKV to the square on KBE or on KE (i.e. z^2) is the same as the ratio of the latus transversum to the latus rectum, i.e. as l to g. Therefore $f^2 - x^2$ will be to z^2 as l is to g, which means that

$$\frac{lz^2}{g} = f^2 - x^2. \qquad [4.59]$$

This however, is the same as $z^2 = f^2 - x^2$, if $l = g$.

And here again it is easy to see that the aforementioned curve will be a circle, if the angle $DKBE$ or DKE is a right angle and $l = g$, which means that the rectangle LKV is equal to the square on KE [3.5].

§2. But if z has been adopted instead of $y \pm (bx / a)$, we first select a point M on line BE

[335]

(produced toward B, if necessary) so that AB is to BM as a is to b, which means in such a way that $BM = bx / a$. This point M however, has to be selected at the same side of the line AB as where the angle ABE has been given or chosen if z has been adopted as $y - (bx / a)$; but if, on the other hand z has been adopted instead of $y + (bx / a)$, it (point M, transl.) has to be selected on the opposite side of the same line AB.

Next we have to draw the straight line $NAMG$ through the points A and M intersecting the straight lines HCH and QFQ at G and N. These lines HCH and QFQ have been drawn through the aforementioned points C and F parallel to BE. After these preparations the diameter of the required ellipse will be situated on the straight line NG so that the lines that are ordinate-wise applied to this diameter include angles equal to the angle AME or to the angle $AMBE$.

Furthermore its center will be at point A and half its latus transversum will be the segment AN or AG. This AN or AG, however, will be equal to ef / a, if we suppose that the ratio of AB to AM is as a to e, because AC, i.e. f, is to AG as AB is to AM or as a is to e.

Finally, the ratio of the latus transversum to the latus rectum will be as $e^2 l$ to $a^2 g$, i.e. as e^2 to a^2

L I B. II. C A P. IV.   335

productâ versùs B, si opùs fuerit, punĉto M; ita ut A B ad B M sit, sicut *a* ad *b*, hoc est, ut B M sit $\infty \frac{b\,x}{a}$, (quod quidem punĉtum M, si *z* assumpta fuerit pro $y - \frac{b\,x}{a}$, ab eadem parte lineæ A B, quâ datus vel conceptus est angulus A B E, sumi debet; sin contra, *z* pro $y + \frac{b\,x}{a}$ assumpta fuerit, ab altera ejusdem lineæ A B parte sumendum est) oportet per punĉta 'A & M rectam lineam ducere N A M G, secantem rectam H C H, atque occurrentem ipsi Q F Q, quæ per prædicta punĉta C & F ipsi B E ductæ sunt æquidistantes, in G & N. Quo facto, erit quæsitæ Ellipseos diameter

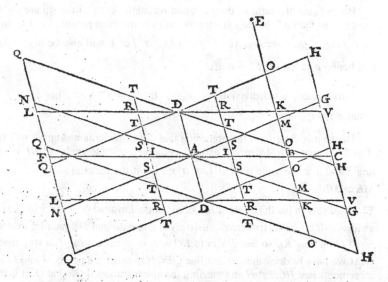

in recta N G, ita ut ad eandem diametrum ordinatim applicatæ cum ea angulos faciant, angulo A M E vel A M B E æquales. Porrò centrum ejusdem erit in punĉto A, & semi-latus transversum erit recta A N vel A G. (quæ quidem A N vel A G, si ratio A B ad A M supponatur ut *a* ad *e*, æquabitur $\frac{e f}{a}$: cum sit ut A B ad A M, sive ut *a* ad *e*, ita A C, hoc est, *f*, ad A G.) Denique ratio transversi lateris ad rectum erit ut *e e l* ad *a a g*, id est, si *l* sit ∞g.

[336]

or as the square on AM to the square on AB, if $l = g$, or, which is the same, if the term z^2 is without a fraction.

Indeed [4.60], if we understand the aforementioned ellipse to be described passing through N and G and if we suppose that it intersects the straight line ME or MBE (ordinate-wise applied to the aforementioned diameter) at point E, then

$$ME = y - \frac{bx}{a} \text{ and } MBE = y + \frac{bx}{a}$$

and so exactly the quantity that has been adopted as z.

Because $AM = ex/a$, therefore $NM = (ef/a) + (ex/a)$ and $MG = (ef/a) - (ex/a)$, and hence the rectangle NMG will be equal to $(e^2 f^2 / a^2) - (e^2 x^2 / a^2)$.

But because the ratio of the aforesaid rectangle NMG to the square on MBE or ME is the same as the ratio of the latus transversum to the latus rectum, i.e. the same as the ratio of $e^2 l$ to $a^2 g$: therefore, $(e^2 f^2 / a^2) - (e^2 x^2 / a^2)$ will also be to z^2 as $e^2 l$ to $a^2 g$ and hence $e^2 l z^2 = e^2 g f^2 - e^2 g x^2$.

This means, after division by $e^2 g$, $l z^2 / g = f^2 - x^2$, or $z^2 = f^2 - x^2$, if we assume $l = g$.

Hence it is clear from the preceding that the aforementioned curve will be a circle with A as its center with AN or AG as half its diameter, if the angle $AMBE$ or AME is a right angle and if at the same time $e^2 l = a^2 g$, i.e. if the rectangle NMG equals the square on ME or MBE.

§3. Finally, if in the third case, z has been adopted instead of $y \pm c \pm (bx/a)$, we first draw as above $AD = c$ (the text erroneously gives f, transl.) and DK parallel to AB and we select point O on line KE so that DK is to KO as a is to b, i.e. in such a way that $KO = bx/a$. Next we have to draw the straight line $QDOH$ through the points D and O intersecting the aforementioned HCH at H and meeting the aforementioned QFQ at Q. It is clear from what we already more often remarked, however, that the aforementioned point O has to be selected at the same side of the line DK as where the angle DKE has been given or chosen if we have $-bx/a$, but that this point has to be selected on the opposite side of the same line if we have $+bx/a$. After these preparations the diameter of the ellipse to be described will be situated on the aforementioned straight line QDH so that the lines that are

∞g, five, quod idem est, si termino zz nulla adhæreat fractio, ut ee ad aa, hoc est, ut A M quadratum ad quadratum A B.

Etenim si prædicta Ellipsis descripta intelligatur , transiens per N & G, supponaturque eandem secare rectam M E vel M B E, ad prædictam diametrum ordinatim applicatam in puncto E: erit eadem M E $\infty\, y - \frac{bx}{a}$, & M B E $\infty\, y + \frac{bx}{a}$, ac proinde ea ipsa, quæ pro z, assumpta est. Cumque A M sit ∞ $\frac{ex}{a}$, erit N M $\infty\, \frac{ef}{a} + \frac{ex}{a}$, & M G $\infty\, \frac{ef}{a} - \frac{ex}{a}$: ideoque rectangulum N M G $\infty\, \frac{eeff}{aa} - \frac{eexx}{aa}$. At cum eadem sit ratio dicti rectanguli N M G ad quadratum ex M B E vel M E, quæ est lateris transversi ad rectum, hoc est, eadem quæ eel ad aag: erit quoque ut eel ad aag, ita $\frac{eeff - eexx}{aa}$ ad zz, ac proinde $eel zz$ $\infty\, eegff - eegxx$. id est, factâ divisione per eeg, erit $\frac{lzz}{g}$ $\infty\, ff - xx$. sive, positâ $l \infty g$, $zz \infty ff - xx$. Vnde ex ante dictis iterum apparet, quòd si angulus A M B E vel A M E rectus sit, ac simul $eel \infty aag$, hoc est, rectangulum N M G ∞ quadrato ex M E vel M B E, prædictam curvam fore Circulum, cujus centrum sit A, & semi-diameter A N vel A G.

§. 3.   Denique si tertiò z assumpta sit pro $y \,\rotatebox{180}{B}\, c\, B\, \frac{bx}{a}$, ductâ, ut supra, A D ∞f, & D K ipsi A B parallelâ, sumptoque in linea K E puncto O, ita ut D K ad K O sit, sicut a ad b, hoc est, ut K O sit $\infty\, \frac{bx}{a}$: ducenda est per puncta D & O recta Q D O H, secans prædictam H C H in H, atque occurrens præfatæ Q F Q in Q. (constat autem ex iis, quæ jam sæpiùs monita sunt, si habeatur $-\frac{bx}{a}$, prædictum punctum O ab eadem parte lineæ D K, quâ datus vel assumptus est angulus D K E, sumendum esse; at si habeatur $+\frac{bx}{a}$, illud ipsum ab altera ejusdem lineæ parte sumi debere.) Quo facto, erit describendæ Ellipseos diameter in prædicta recta Q D H, ita ut ad eandem diametrum ordinatim applicatæ

plicatæ

[337]

ordinate-wise applied to this diameter enclose angles equal to the angle *DOKE* [4.61] or *DOE*. Furthermore, the center will be at *D* and half the latus transversum *DQ* or *DH* will

equal to *AN* or ef/a and the ratio of the latus transversum to the latus rectum is as $e^2 l$ to

$a^2 g$ [4.62].

Indeed, if we understand the required ellipse to have been described passing through the points *Q* and *H* and if we suppose that it intersects the straight line *OE* or *OKE* at point *E*, then

$$OKBE = y + c + \frac{bx}{a}, \qquad OE = y - c - \frac{bx}{a},$$

$$OBE = y + c - \frac{bx}{a}, \text{ and } OKE = y - c + \frac{bx}{a}$$

and hence the aforementioned lines will be the same as those that have been adopted as *z*. As moreover

$$DO \text{ (or } AM) = \frac{ex}{a} \text{ and so } QO = \frac{ef}{a} + \frac{ex}{a} \text{ and } OH = \frac{ef}{a} - \frac{ex}{a},$$

therefore the rectangle *QOH* will be equal to $(e^2 f^2 - e^2 x^2)/a^2$. Because the ratio of the aforementioned rectangle *QOH*

L I B. II. C A P. IV.   337

plicatæ cum ea angulos faciant , angulo D O K E vel D O E
æquales. Porrò centrum erit in D , & femi-latus tranfverfum
D Q vel D H ∞ A N feu $\frac{ef}{a}$, ac ratio tranfverfi lateris ad rectum,
ut *eel* ad *aag.*

Si enim quæfita Ellipfis defcripta intelligatur , tranfiens per
puncta Q & H, eademque fecare fupponatur rectam O E vel
O K E in puncto E: erit O K B E ∞ $y + c + \frac{bx}{a}$, O E ∞

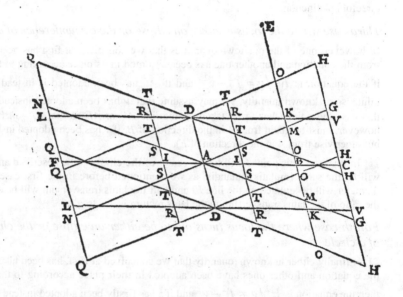

$y - c - \frac{bx}{a}$, O B E ∞ $y + c - \frac{bx}{a}$, & O K E ∞ $y - c +$

$\frac{bx}{a}$: ac proinde prænominatæ illæ lineæ eædem erunt, quæ pro

z affumptæ funt. Cumque porrò fit D O feu A M ∞ $\frac{ex}{a}$, ideoque

Q O ∞ $\frac{ef}{a} + \frac{ex}{a}$, & O H ∞ $\frac{ef}{a} - \frac{ex}{a}$: erit rectangulum Q O H

∞ $\frac{eeff - eexx}{aa}$. At cum eadem fit ratio dicti rectanguli Q O H

Pars II.   V u   ad

[338]

to the square on *OKBE* or *OE* or to the square on *OBE* or *OKE*, is the same as the ratio of the latus transversum to the latus rectum, i.e. as e^2l to a^2g, therefore $(e^2f^2 - e^2x^2)/a^2$ will be to z^2 as e^2l to a^2g and hence $e^2lz^2 = e^2gf^2 - e^2gx^2$ and, after division of all terms by e^2g, $lz^2/g = f^2 - x^2$. This means $z^2 = f^2 - x^2$, if $l = g$.

And here again it is easily seen that the aforementioned curve will be a circle, if the angle *DOKBE, DOE, DOBE* or *DOKE* is a right angle and if at the same time $e^2l = a^2g$.

This, however, is all that was to be demonstrated in the aforementioned case with its threefold distinction.

Third case where the locus is either an ellipse or the circumference of a circle

If, however, one of the unknown quantities that we conceived at first has been taken away from the equation and another one has been assumed in its place according to the Rule, and if the equation is $ly^2/g = f^2 - v^2$, and if v has been adopted is instead of x plus or minus some known quantity, we may assume that v has been adopted instead of $x \pm h$. In this case the point *I* has to be taken on the line *AB* or *AF* so that $AI = h$ (this point *I*, however, must be taken from *A* in the direction of *B*, if v has been adopted instead of $x - h$, but otherwise from *A* in the direction of *F*).

In this way, this point *I* will be the center of the ellipse to be described and all the rest will be the same (mutatis mutandis) as we mentioned before in the first case. This is, the diameter will be situated on the line *IB* and half the latus transversum will be equal to *f* and the ratio of the latus transversum to the latus rectum is as *l* to *g*.

Fourth case where the locus turns out to be either an ellipse or the circumference of a circle

§1. If finally either unknown quantity that we conceived at first, has been taken away from the equation and other ones have been adopted in their place according to the Rule and if then the equation is $lz^2/g = f^2 - v^2$ and if *z* has firstly been adopted instead of $y \pm c$, then we have to draw *IR* on either side parallel to *BE* and equal to *c*. Then this same point *R* will be the center of the ellipse and its diameter will be situated on the straight line *RK* or *RL* and half its latus transversum will be equal to *f* and the ratio of the latus transversum to the latus rectum will be as *l* to *g*, just as all that at greater length has been explained in §1 of the second case with the necessary changes having been made (p. [334], transl.).

§2. But if *z* has been adopted instead of $y \pm (bx/a)$, the point *S*, at which *MA*,

338 ELEM. CVRVARVM

ad quadratum ex $OKBE$ vel OE, aut ad quadratum ex OBE vel OKE, quæ est transversi lateris ad rectum, hoc est, ut eel ad $a\,ag$: erit quoque ut eel ad $a\,ag$, ita $\frac{eeff-eexx}{aa}$ ad zz; ac propterea $eelzz \infty eegff - eegxx$, &, divisis omnibus per eeg, $\frac{lzz}{g} \infty ff - xx$. id est, si l sit ∞g, erit $zz \infty ff - xx$.

Atque hîc iterum facilè apparet, si angulus $DOKBE, DOE$, $DOBE$, vel $DOKE$ rectus foret, & simul $eel \infty aag$, prædictam curvam fore Circulum. Quæ quidem omnia sunt, quæ supra dicto casu in triplici sua variatione demonstranda erant.

Casus 3^{tius}, cùm Locus est vel Ellipsis vel Circuli circumferentia. Si verò quantitatum incognitarum ab initio conceptarum alterâ ex æquatione sublatâ, aliâque ejusdem loco secundùm Regulam assumptâ, æquatio sit $\frac{lyy}{g} \infty ff - vv$, atque ipsa v assumpta sit pro x ☖ notâ aliquâ quantitate; Sit v assumpta pro x ☖ h, eritque eo casu in linea $A\,B$ vel $A\,F$ sumendum punctum I; ita ut $A\,I$ sit ∞h. (quod quidem punctum I, si v assumpta fuerit pro $x - h$, ab A versùs B; sin contra ab A versùs F sumi debet.) Quo facto, erit idem punctum I centrum describendæ Ellipseos, &, mutatis mutandis, cætera omnia, ut supra, casu primo memoratum est. Hoc est, diameter erit in recta IB, ac semi-latus transversum erit ∞f, atque ratio transversi lateris ad rectum, ut l ad g.

Casus 4^{ttus}, cùm Locus vel Ellipsis vel Circuli circumferentia existit.

§. 1. Si denique quantitatum incognitarum primùm conceptarum utrâque ex æquatione sublatâ, aliisque earundem loco juxta Regulam assumptis, æquatio sit $\frac{lzz}{g} \infty ff - vv$; atque z primò assumpta sit pro y ☖ c, ducenda est utrinque IR, parallela ipsi BE, ac ∞c. Quo facto, erit idem punctum R centrum Ellipseos, & diameter ejus in recta RK vel RL, eritque ejus semilatus transversum ∞f, ac ratio transversi lateris ad rectum, ut l ad g. quemadmodum ea omnia Casu 2do §. 1, mutatis mutandis, fusiùs explicata sunt.

§. 2. At si z assumpta fuerit pro y ☖ $\frac{bx}{a}$, erit punctum S, ubi MA,

vel,

[339]

or its conjugate [4.48], is intersected by the aforementioned *IR* (produced, if necessary) will be the center of the ellipse, and all the rest will work out as we explained before in §2 of the second case, with the necessary changes having been made. Indeed, the diameter of the conic section will be situated on *SM* (and such that instead of $AM = ex/a$, here we have $SM = ev/a$, because *BI*, or *v*, is to *SM* as *BA* is to *AM*, i.e. as *a* is to *e*). Furthermore half the latus transversum will be ef/a and the ratio of the latus transversum to the latus rectum will be as e^2l to a^2g.

§3. If, finally, *z* has been adopted instead of $y \pm c \pm (bx/a)$, point *T* at which *DO,* or its conjugate [4.48], is intersected by the aforementioned *IR*, (produced if necessary) will be the center of the ellipse; and the rest will work out just as we explained at greater length in the preceding section and above in §3 of the second case (p. [336], transl.).

vel, quæ ipfi in directum adjungitur, per prædictam I R, productam, fi opùs fuerit, interfecatur, centrum Ellipfeos; & cætera omnia, mutatis mutandis, ut fupra cafu fecundo §. 2. memoratum eft. Nempe erit fectionis diameter in recta S M, (atque ut ibidem erat A M $\infty \frac{ex}{a}$, ita hîc S M erit $\frac{cv}{a}$: cum fit ut B A ad A M, hoc eft, ut a ad e, ita B I, id eft, v, ad S M:) eritque porrò femi-latus tranf-

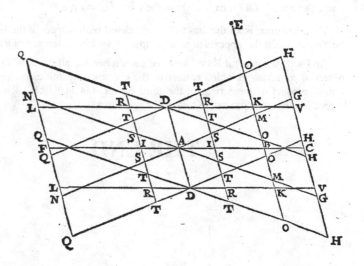

verfum $\infty \frac{ef}{a}$, & ratio tranfverfi lateris ad rectum, ut eel ad aag.

§. 3.   Denique fi z affumpta fuerit pro y $\backslash \!\! 8$ c $\backslash \!\! 8$ $\frac{bx}{a}$, erit punctum T, in quo D O, vel, quæ ipfi in directum adjungitur, per prædictam I R, productam, fi opùs fuerit, interfecatur, centrum Ellipfeos; & reliqua omnia, mutatis mutandis, ut paragrapho præcedenti ac fupra cafu fecundo §. 3

[340]

Indeed, the diameter will be situated on the straight line TO and half the latus transversum will be equal to ef/a and the ratio of the latus transversum to the latus rectum will be as e^2l to a^2g.

The demonstration of all these statements has clearly been explained in the preceding, while all terms and quantities here correspond to the former ones with only one exception: the quantities that there were denoted by x, are here denoted by $x \pm h$, that is by v.

Indeed where there was "$AB = x$," here there is "$IB = v$"; where there was "$DK = x$," here there is "$RK = v$"; where there was "$AM = ex/a$," here there is "$SM = ev/a$"; and where there was "$DO = ex/a$," here there is "$TO = ev/a$."

This, however, is all that has to be considered with respect to the finding of a locus in the case in which this appears to be an ellipse or the circumference of a circle.

And so, by a general Rule, we have comprehended all cases of the finding of loci by means of equations in which neither of the unknown quantities multiplied by itself nor a mutual product of them rises to the third power, but in which such a product does not exceed a square or a product of two different factors.

THE END

340   ELEM. CVRVAR. LIB.II. CAP. IV.

fusiùs explicatum est. Nempe erit diameter in recta T O , & semi-latus transversum $\infty \frac{ef}{a}$, ac ratio transversi lateris ad rectum, ut $e\,e\,l$ ad $a\,a\,g$. Atque eorum omnium demonstratio in præcedentibus explicitè est comprehensa , cum termini & quantitates omnes hîc cum prioribus conveniant; excepto tantùm , quòd quæ ibidem designabantur per x hîc designentur per x ౸ h , hoc est , v. Ita enim quòd ibi erat A B ∞ x, hîc est I B ∞ v; quod ibi erat D K ∞ x, hîc est R K ∞ v ; quod ibi erat A M ∞ $\frac{ex}{a}$, hîc est S M ∞ $\frac{ev}{a}$; & quod ibi erat D O ∞ $\frac{ex}{a}$, hic est T O ∞ $\frac{ev}{a}$.

Quæ quidem omnia sunt, quæ circa inventionem Loci illo casu, quo idem vel Ellipsis vel Circuli circumferentia existit, consideranda veniunt.

Atque ita generali Regulâ casus omnes inveniendi Loca per æquationes, in quibus neutra quantitatum incognitarum in se ducta nec factum sub iisdem ad tres dimensiones ascendit, sed vel quadratum vel planum non excedit, complexi sumus.

F I N I S.

4

Annotations to the translation

Chapter I

[1.1] What is meant here is the equation that determines the relation between the abscissa x and the ordinate y of an arbitrary point on the curve (the locus). Incidentally, this introduction reminds us of the opening sentence of the *Géométrie* of Descartes:

> *Tous les Problesmes de Geometrie se peuvent facilement reduire a tels termes, qu'il n'est besoin par aprés que de connoistre la longeur de quelques lignes droites, pour les construire.* [All geometry problems can easily be reduced to such terms that to construct them only the lengths of certain line segments are needed.]

[1.2] The *unknown and indeterminate line segments* designate the abscissa and the ordinate of an arbitrary point on the curve in question. In the computations these are treated as if they were known, which leads to an equation; this is typical for the analytic method (see Introduction).

Jan de Witt uses almost exactly the same words as Pappus (Ἀνάλυομενος, vii):

> Analysis is a method where one assumes that which is sought to be known and from that arrives, through a chain of logical deductions, at something that is recognized as true.

[1.3] The plane loci are the straight line and circle; the solid loci are the conics parabola, hyperbola, and ellipse (no circle). See also Note [1.20] to *Liber Primus* (pp. [219] and [220]).

[1.4] For the terms *determinatio* (determination), *descriptio* (definition), and *demonstratio* (demonstration), see Introduction and Summary.

[1.5] As the quantities are strictly positive, the case $y = -(bx/a) - c$ is not considered.

[1.6] The quantities x and y are clearly line segments. When Jan de Witt writes that by assumption x *extends indefinitely along a straight line whose position is given* (*ex certo et immutabili illo initio in linea recta positione data intelligatur indefintie extendi*), he means that it is a finite line segment whose length can be chosen to be arbitrarily large. In particular, the line segments stay finite and do not become

A.W. Grootendorst et al. (eds.), *Jan de Witt's Elementa Curvarum Linearum*, Sources and Studies in the History of Mathematics and Physical Sciences, DOI 10.1007/978-0-85729-142-4_4, © Springer-Verlag London Limited 2010

half-lines going on infinitely in one direction. This is how the recurring sentence that *x extends indefinitely along the line AB* should be interpreted.

[1.7] *Immutabile* is associated with *initium* and not with *punctum A*. This can be deduced from p. [244], lines 14 and 15 from the top: *ut initium unius, verbi gratia, ipsius x, certum sit et immutabile*, and from p. [267], line 4 from the bottom, where we read: *Sit initium immutabile ipsius x punctum A*.

Incidentally, we can also think of the flux theory of Aristotle, in which he considers a line to be the path of a point (*Physica* **262**).

[1.8] Here, as in so many other places, a letter, in this case *C*, does not denote a fixed point but a variable one.

[1.9] Jan de Witt quotes the *Elements* VI, 16, which shows how geometric his thinking was. It states:

> if four line segments are proportional, then the *rectangle* formed by the outer terms is equal to the *rectangle* formed by the inner terms.

[1.10] The procedure followed here is typical. Jan de Witt starts with an equation, here $y = bx/a$, and uses the parameters a and b that occur in it to construct a curve, in this case a straight line (*descriptio*). He then takes an arbitrary point on it and shows that the abscissa and ordinate of that point satisfy the equation he started out with (*demonstratio*). He does not, however, show that every point whose abscissa and ordinate satisfy the equation lies on this curve. Because of this he later misses the second branch of a hyperbola.

[1.11] The text erroneously states $ED = a + c$.

Chapter II

[2.1] This refers to the *Regula Universalis* on p. [255].

[2.2] In the figure on p. [249] and the cases that follow this latus rectum is represented by the line segment tangent to the parabola at its vertex. This accentuates the geometric viewpoint: for example in the figure that was just mentioned, we can now see the statement " $y^2 = ax$ " as $DE^2 = FA \cdot AE$, that is, as *"the area of the square on DE is equal to the area of the rectangle with sides FA and AE."*

For the term "latus rectum" see also *Liber Primus* p. [164] and Appendix B, p. 277.

[2.3] On p. [247], line 14 from the top, Jan de Witt uses the term *ad binos rectos complemento* instead of *supplemento*.

[2.4] Again it is clear that Jan de Witt only considers that part of the curve whose points have a "positive" abscissa.

[2.5] The four cases are those mentioned on p. [249]. The method stated here amounts to "splitting off a square." Indeed if, for example, $x^2 \pm 2xy$ (that is to say, $x^2 \pm x \cdot 2y$) comes up, then Jan de Witt prescribes the substitution $z = x \pm y$.

[2.6] Strictly speaking, the fact that the equation $z^2 = dx$ has the same form as the equation $y^2 = dx$ in Theorem VII does not guarantee that this also represents a parabola; this, of course, depends on the relations between z and x and y. If for example $z = y^2$, then the conclusion obviously does not hold.

[2.7] It is clear that this d is not the latus rectum of the parabola in question with equation $z^2 = dx$. Indeed it concerns the parabola passing through D with vertex G and diameter GC. This is explained on p. [259]; as it comes back frequently in this book, we will summarize this reasoning here.

If we let p denote the latus rectum associated with the diameter GC, then (see the figure on p. [258]) $DC^2 = p\,GC$. Setting $GB : BC : CG = a : b : e$, this gives

$$GC = \frac{ex}{a} \text{ (as } GB=x\text{), and, therefore, } DC^2 = p\,\frac{ex}{a}.$$

Moreover, $DC^2 = dx$, hence $p = ad/e$.

In the figure on p. [258] this latus rectum p is represented by the line segment GF. The point F is defined explicitly on p. [259], line 9 from the top.

[2.8] Again we see that Jan de Witt only accepts those parts of curves that lie above the abscissa-axis: that is, above the half-line to the right of A.

[2.9] Jan de Witt will now show in the usual way that the abscissa and ordinate of an arbitrary point on the curve that he has constructed "intentionally" satisfy the equation he started out with. Again the *demonstratio* follows the *descriptio*.

[2.10] Jan de Witt speaks of *the* diameter, which, though somewhat sloppy, is clear enough.

[2.11] This concerns the equation $dy = v^2$ of p. [260], line 3 from the bottom.

[2.12] This concerns the figure on p. [261].

[2.13] According to p. [262], line 12 from the top, $GC = ey/a$; according to line 16 from the top, the latus rectum FG is chosen equal to ad/e.

As $DC = v$ and $DC^2 = FG \cdot GC$, this gives $dy = v^2$.

[2.14] See p. [305].

[2.15] See also Note [2.7]. The parabola described here is not yet the final result; the term d^2 in the equation must still be taken into account. Jan de Witt achieves this by translating the vertex A along BAG from A to G, so that

$$z^2 = DB^2 = FG \cdot GB = FG \cdot AB + FG \cdot GA.$$

We already saw that

$$FG \cdot AB = \frac{2ab}{e}\,\frac{ex}{2a} = bx.$$

To conclude we need $FG \cdot GA = d^2$, hence $2(ab/e)GA = d^2$, and therefore

$$GA = \frac{d^2 e}{2ab}, \text{ giving } AC = \frac{d^2}{b}.$$

[2.16] A is translated along EAC from A to C, so that $AC = d^2/b$. The proof is clear because $CA : GA = AE : AB = 2a : e$ (p. [264], line 5 from the bottom); if moreover $AC = d^2/b$, then GA has the desired value $(d^2/b) \cdot (e/2a) = d^2 e/2ab$.

[2.17] This is the case obtained by interchanging x and y.

[2.18] It has been tacitly assumed that CD is parallel to AE.

[2.19] The reasoning is as follows (see the figure on p. [266] for the meaning of the letters used here): if we let p (= FG) denote the latus rectum associated with the diameter BKG, then $BD^2 = FG \cdot GB$, hence $v^2 = FG \cdot GB$, and therefore $v^2 = p(BK + KG) = p((ey/a) + KG)$.

We must also have $v^2 = by + (c^2/2)$, hence $by = p(ey/a)$, whence $p = ab/e$ and $p.KG = c^2/2$. Consequently $KG = c^2 e/2ab$ and

$$GB = GK + KB = \frac{c^2 e}{2ab} + \frac{ey}{a}.$$

We now know the following for the parabola we have in mind: a diameter with associated latus rectum, the vertex, and the conjugate direction.

[2.20] *Determinare* and *describere* are used as synonyms.

[2.21] See also Note [1.7].

[2.22] This is the case obtained by interchanging x and y.

[2.23] See the illustration on p. [269]. This is where the parabola we have in mind is determined (*determinatio*).

[2.24] The reasoning is as follows (see the figure on p. [270] for the meaning of the letters used here): Jan de Witt is looking for a parabola corresponding to the equation

$$x^2 - \frac{2bxy}{a} + \frac{b^2 y^2}{a^2} + dy - c^2 = 0,$$

which he has reduced to the form $v^2 = c^2 - dy$, and tries the parabola with diameter GA, conjugate direction AE, and vertex G. An arbitrary point D on this must satisfy $BD^2 = p \cdot GB$, where p (= GF) is the latus rectum, which has not been fixed yet. Now C and G are chosen in such a way that $AC : CG : GA = a : b : e$ and $AC = c^2/d$, and hence

$$BD = HD - HB = x - \frac{by}{a} = v$$

and

$$GB = GA - BA = \frac{c^2 e}{ad} - \frac{ey}{a}.$$

We have $BD^2 = GF \cdot GB$; that is:

$$v^2 = p\left(\frac{e}{ad}\right)(c^2 - dy).$$

Moreover, $v^2 = c^2 - dy$, so that $pe/ad = 1$ and $p = ad/e$.

This p appears more or less out of nowhere in the *descriptio* on p. [271], lines 6 and 7 from the bottom, as does the choice of AC, but in the *demonstratio* everything of course turns out right.

[2.25] The reasoning is analogous to that in Note [2.24].

[2.26] See Chapter IV, p. [304].

[2.27] It follows clearly from the context that "$y = a$" means the absolute value of $y - a$.

In A *History of Mathematical Notations* (Part I, pp. 158 and 183), Cajori mentions the use of this notation by Viète (1540–1603) and by Albert Girard (1595–1632). Viète calls the "=" sign the *minus incertus*. Girard says:

> *Touchant les lettres de l'Alphabet au lieux des nombres: soit A & aussi B deux grandeurs: ... leur différence est $A = B$ (ou bien si A est majeur on dira que c'est $A - B$)* [When using the letters of the alphabet instead of numbers: let A and B be two quantities: ... their difference is $A = B$ (or if A is greater than B, we will say that it is $A - B$)] and uses the "=" sign as *difference entre deux quantitez ou il se treuve* [*difference of the two quantities between which it is placed*].

The two vertical lines that we use to denote the absolute value were introduced in 1841 by Karl Weierstraß (1815–1897).

[2.28] Computations show that the given point A lies on the symmetry axis of the described parabola, which has vertex H, in such a way that AH is one fourth of the latus rectum of the parabola in question. "Therefore," says Jan de Witt, "A is exactly the point that is called the focus or umbilicus." The term "focus" was introduced by Kepler in 1604 (*Collected Works* II, p. 91).

The notion of "focus" was already known in the Greek antiquity. It is remarkable that in his *Conica*, Apollonius mentions the foci of the hyperbola and the ellipse (see also Notes [3.26] and [3.42]), but not that of the parabola. We can only guess why this is the case.

According to Toomer [58] the focus of a parabola had already been mentioned around 250 BC by Dositheus, a friend of Archimedes, as the point through which all rays parallel to the axis of a parabola pass after reflection in the parabola. In his work *On burning mirrors*, Diocles of Carystus, a contemporary of Apollonius, gives a correct geometric proof of the theorem that the point inside the parabola that lies on its symmetry axis at a distance from the vertex equal to one-fourth of the latus rectum satisfies the property characterizing the focus. This work is only known in its Arabic translation and is supposed to have been called περί πυρίων.

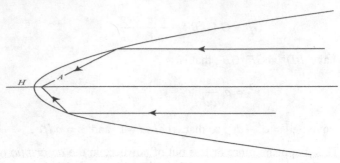

FIGURE 4.1

For this see the work *Diocles on burning mirrors* of Toomer mentioned above.

[2.29] The axis only reaches to the point *H*.

[2.30] In *Liber Primus*, Corollary 2 (not 1) on p. [174], Jan de Witt deduces the theorem

> ... that arbitrary straight lines touching a parabola, wherever it may be, and lines that are ordinate-wise applied to the diameter from the point of contact, cut off equal parts of the diameter on both sides of the vertex and vice versa...

[2.31] Even for this simple conclusion Jan de Witt refers to the *Elements* of Euclid (Book I, Theorem 5).

Chapter III

[3.1]   The reference is to p. [243], lines 1–5 from the bottom, where we read:

> If, however, each of the two unknown quantities has been raised to the second power or if a mutual product occurs in the equation, then the required locus will be a hyperbola or an ellipse or a circle.

As Jan de Witt has already stated shortly prior to that on p. [243], the equations in question have been reduced to the simplest form; we would say that they have been "reduced to the standard form." Otherwise the statement would of course be false (see also his examples of a parabola on pp. [263] and [267]). He repeats this here on p. [275].

In the second *Regula Universalis* (p. [280]), he restricts himself from the beginning to hyperbolas, ellipses, and circles.

[3.2]   These cases I–IV are treated in almost the same manner in Chapter IV, but then as preliminaries to situations where *x* and *y* are replaced by linear combinations of *x* and *y*.

[3.3]   If $l = g$, the equation is $y^2 = x^2 - f^2$. By choosing the latus rectum *GF* equal to the transverse diameter *CG*, Jan de Witt ensures that this equality holds.

For the construction of the hyperbola he uses the instructions of *Liber Primus*, Chapter IV, pp. [231]–[235].

Theorem IX, Proposition 10 of *Liber Primus* (p. [196]) is of fundamental importance here and in many other places in the book where hyperbolas are concerned. We recall that theorem here.

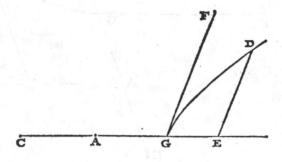

FIGURE 4.2.

When the transverse diameter CG and the conjugate diameter are denoted by $2a$ and $2b$, respectively, then according to the theorem, the following holds in the figure on p. [277] of *Liber Secundus* (= Figure 4.2):

$$DE^2 : CE \cdot GE = 4b^2 : 4a^2 .$$

The latus rectum p (= GF) is defined by $2a : 2b = 2b : p$, so that

$$DE^2 : CE \cdot GE = 4b^2 : 4a^2 = p : 2a = GF : CG .$$

In the situation on p. [276] (the first case) Jan de Witt deliberately chooses

$$CG = GF , \text{ so that } DE^2 : CE \cdot GE .$$

It now follows from the construction that by taking $a = f$, we also have

$$DE = y, CE = AE + CA = x + f, \text{ and } GE = AE - AG = x - f,$$

whence the coordinates of D satisfy $y^2 = x^2 - f^2 .$

If $l \neq g$, Jan de Witt chooses GF and CG in such a way that $GF : CG = g : l$, so that

$$y^2 : (x^2 - f^2) = p : 2a = GF : CG = g : l$$

and therefore $ly^2 / g = x^2 - f^2$, which is the situation on p. [277].

[3.4] This is just the previous theorem with x and y interchanged. It concerns the figure on p. [278] (= Figure 4.3).

[3.5] Theorem XII, Proposition 13 of *Liber Primus* (p. [205]) is of fundamental importance here and in many other places in the book where ellipses are concerned. We recall it here:

When the transverse diameter and the conjugate diameter are denoted by $2a$ and $2b$, respectively, then according to the theorem, the following holds in the figure on p. [279] (= Figure 4.4; note that DE is not necessarily perpendicular to CG):

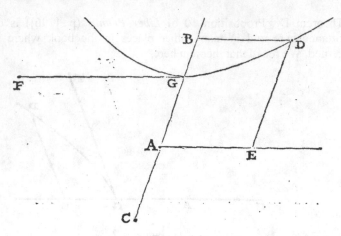

FIGURE 4.3

$$DE^2 : CE \cdot GE = 4b^2 : 4a^2 .$$

The latus rectum p $(=GF)$ is again defined by $2a : 2b = 2b : p$, so that

$$DE^2 : CE \cdot GE = 4b^2 : 4a^2 = p : 2a = GF : CG.$$

From the construction also follows that

$$DE = y, \ CE = CA + AE = f + x, \text{ and } GE = AG - AE = f - x.$$

Therefore, for the given choice of $GF : CG$, the coordinates of D satisfy

$$y^2 : (f^2 - x^2) = p : 2a = GF : CG = g : l ,$$

so that $\qquad \dfrac{ly^2}{g} = f^2 - x^2 .$

If $2a = 2b$, then $l = g$ and $x^2 + y^2 = f^2$. If moreover DE is perpendicular to CG, then

$$AD^2 = f^2 = \text{constant, in which case we have a circle.}$$

[3.6]   The four cases that this general Rule refers to are those on p. [275]. The difference between this general Rule and that of p. [255] lies in the nature of the curves under consideration. On p. [255] only the parabola is discussed, in the four forms of p. [249]; here, on p. [275], the hyperbola and the ellipse (or the circle) are concerned.

This difference is reflected in the equations that are considered. Jan de Witt mentions the following terms explicitly, alone or in combination with each other: xy, x^2, y^2, ax, ay, and axy/b , where a and b are known quantities. A term such as axy cannot occur, of course, because of the homogeneity condition.

He gives the following two substitution rules:

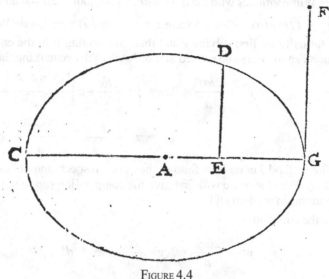

FIGURE 4.4

(i) If the combination $xy \pm ay$ occurs and we wish to replace x, then we set $v = x \pm a$. In the description of this substitution with words, x is the quantity that is to be replaced, which is not squared, and a is the coefficient of a quantity that is not to be replaced. We would say "put y outside of the brackets," but that is an anachronism. For the origin of "brackets," we refer the reader to Cajori's *A History of Mathematical Notations*, pp. 342–355.

(ii) If the combinations $x^2 \pm ax$, $x^2 \pm xy$, or $x^2 \pm ay/b$ occur, where the quantity that is to be replaced –here x– is squared, then the substitutions $v = x \pm a/2$, $v = x \pm y/2$, or $v = x \pm ay/2b$ respectively are prescribed, which amount to "splitting off a square."

What has been said about x in (i) and (ii) also holds mutatis mutandis for y.

The general Rule now says that after (possibly repeated) application of the substitutions given here one always arrives at one of the cases of p. [275]; that is to say, at a hyperbola or an ellipse (or a circle), at least if the curve in question is not a parabola.

In Chapter IV the cases of p. [275] are generalized by replacing x and y by linear combinations of x and y.

This Rule is not proven, but is demonstrated by means of a number of examples taking up the rest of Chapter III.

[3.7] See also the beginning of Note [2.6].

[3.8] The terms *determinatio* and *descriptio* are used as synonyms.

[3.9] The use of the Greek genitivus pluralis asymptot ω n with the Greek letter ω is worth noting.

[3.10] Jan de Witt continues with the *demonstratio* without the usual announcement.

[3.11] In $y^2 + (2bxy/a) + 2cy$ we have $z = y + (bx/a) + c$. Jan de Witt goes to work very formally by first solving y and then substituting it in the equation. The final solution can of course be arrived at more quickly by remarking that

$$y^2 + \frac{2bxy}{a} + 2cy = \left(y + \frac{bx}{a} + c\right)^2 - \frac{b^2x^2}{a^2} - \frac{2bcx}{a} - c^2$$

$$= z^2 - \frac{b^2x^2}{a^2} - \frac{2bcx}{a} - c^2 \,.$$

The method Jan de Witt follows, however, respects the indications he gave on p. [281]. For clarity we will first give the computation (up to p. [286], line 4 from the bottom) in modern style.

In the equation

$$y^2 + \frac{2bxy}{a} + 2cy = \frac{fx^2}{a} + ex + d^2 \,,$$

the substitution $z = y + (bx/a) + c$ gives

$$z^2 - \frac{b^2x^2}{a^2} - \frac{2bcx}{a} - c^2 = \frac{fx^2}{a} + ex + d^2 \,,$$

that is: $$\frac{a^2z^2}{af + b^2} = x^2 + \frac{a^2ex + 2abcx}{af + b^2} + \frac{a^2d^2 + a^2c^2}{af + b^2} \,.$$

If we set $$v = x + \frac{a^2e + 2abc}{2af + 2b^2}$$

and $$2h = \frac{a^2e + 2abc}{af + b^2} \,,$$

so that $v = x + h,$

then we find $$\frac{a^2z^2}{af + b^2} = v^2 - \left\{h^2 - \frac{a^2d^2 + a^2c^2}{af + b^2}\right\}.$$

This leads to the distinction of two cases:

(i) $h^2 > \dfrac{a^2d^2 + a^2c^2}{af + b^2}$;   (ii) $h^2 < \dfrac{a^2d^2 + a^2c^2}{af + b^2}$.

In both cases this is a hyperbola; the difference comes from the interchange of the abscissa and the ordinate. The second case is discussed from p. [286], line 3 from the bottom, to p. [288].

The degenerate case, $h^2 = (a^2d^2 + a^2c^2)/(af + b^2)$, is not considered.

Jan de Witt now concludes that the curve in question is a hyperbola and tries to transform it to the hyperbola drawn in the figure on p. [284], which we repeat here as Figure 4.5.

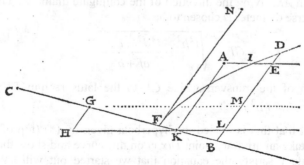

FIGURE 4.5

First the equation is reduced to the form

$$\frac{i^2 z^2}{af + b^2} = \frac{i^2 v^2}{a^2} - \left(\frac{i^2 h^2}{a^2} - \frac{i^2 d^2 + i^2 c^2}{af + b^2} \right),$$ (*)

where i is determined by the proportion $KH : HG : GK = a : b : i$.

Jan de Witt now claims that this equation represents a hyperbola. Moreover, he has a special hyperbola in mind for it. This hyperbola is constructed using the coefficients a, b, c, d, e, and f that occur as follows in the given equation.

AK is drawn downward through A under the "given angle" with AE and with length c. Then KL is drawn parallel to AE; on this a segment $KH = h$ is set out from K, to the left. If H is chosen as origin and HK as abscissa-axis, then this corresponds to the substitution $v = x + h$. After this a line is drawn through H, parallel to AK, on which a point G is chosen so that $GH : HK = b : a$. Next the line GK is drawn. The ratio $KG : KH$ is now also fixed; we let it equal $i : a$, giving $KH : HG : KG = a : b : i$ and therefore also $KL : LB : KB = a : b : i$. We now choose GK as abscissa-axis, G as origin on it, and denote the abscissa by w and the ordinate by z. This construction corresponds to the substitutions

$$w = \frac{i(x + h)}{a} = \frac{iv}{a} \text{ and } z = y + \frac{bx}{a} + c.$$

In spite of this, Jan de Witt does not introduce a w, but continues working with iv/a, a term which he notes should, after all multiplications and divisions, occur squared with coefficient 1, hence as $i^2 v^2/a^2$ "sec." After all, he is working towards the equation of Theorem XII; that is:

$$\frac{ly^2}{g} = x^2 - f^2,$$

where in our case y has been replaced by z and x by $w = iv/a$.

The hyperbola in question now has G as center and GK as support of the diameter, with AK giving the direction of the conjugate diameter. The half-length of the transverse diameter is chosen to be

$$CG = GF = \sqrt{\frac{i^2h^2}{a^2} - \frac{i^2d^2 + i^2c^2}{af + b^2}},$$

and the ratio of the transversal side CF to the latus rectum FN is set to be $CF : FN = i^2 : (af + b^2)$.

The proof that this is indeed the hyperbola given by (*) (top of this page) is simple. We take an arbitrary point $D(x,y)$ on this curve and show that its abscissa x and ordinate y satisfy the equation that we started out with. We first draw $DELB \parallel AK$. From Figure 4.5 then follows that

$$KL : LB : KB = a : b : i, \text{ and hence } LB = \frac{bx}{a},$$

so that

$$DB = DE + EL + LB = y + c + \frac{bx}{a} = z.$$

We also have $GK : HK = i : a$,

and therefore $GK = \dfrac{ih}{a}$ and $KB = \dfrac{ix}{a}$,

so that

$$GB = GK + KB = \frac{i(x+h)}{a} = \frac{iv}{a}.$$

By the characteristic property of a hyperbola, we also have

$$DB^2 : BC \cdot BF = FN : CF.$$

As

$$DB = y + c + \frac{bx}{a} = z,$$

and therefore

$$z^2 : \left(\frac{iv}{a} + CG\right)\left(\frac{iv}{a} - GF\right) = FN : CF = (af + b^2) : i^2,$$

so that

$$z^2 : \left(\frac{i^2v^2}{a^2} - \frac{i^2h^2}{a^2} + \frac{i^2d^2 + i^2c^2}{af + b^2}\right) = \left(af + b^2\right) : i^2,$$

we have

$$\frac{i^2z^2}{af + b^2} = \frac{i^2v^2}{a^2} - \frac{i^2h^2}{a^2} + \frac{i^2d^2 + i^2c^2}{af + b^2}.$$

If we set

$$v = x + h, z = y + \frac{bx}{a} + c,$$

and

$$2h = \frac{ea^2 + 2bca}{af + b^2},$$

the equation of this hyperbola turns out to be the given equation

$$y^2 + \frac{2bxy}{a} + 2cy = \frac{fx^2}{a} + ex + d^2.$$

The case

$$h^2 < \frac{a^2d^2 + a^2c^2}{af + b^2}$$

is analogous and also gives a hyperbola, where the abscissa and ordinate have been interchanged (see the figure on p. [287]).

[3.12] This is shown using the substitution that follows right after it.

[3.13] This converse is considered on p. [286], line 3 from the bottom (in the Latin text).

[3.14] This concerns the form $ly^2/g = x^2 - f^2$. If we have abscissa $GB = w = iv/a$ and ordinate $DB = z$, then the equation is

$$\frac{lz^2}{g^2} = \left(\frac{iv}{a}\right)^2 - f^2.$$

[3.15] Jan de Witt starts out with the equation in the form

$$\frac{a^2z^2}{af + b^2} = v^2 - h^2 + \frac{a^2d^2 + a^2c^2}{af + b^2}, \tag{*}$$

where $z = y + (bx/a) + c$ and $v = x + h$. See p. [283], lines 9 and 10 from the bottom. In the equation that he wishes to arrive at, where the line through C, G, F, and K plays the role of the v-axis, $GB^2 (= (iv/a)^2)$ must have coefficient 1 in order to obtain the form given in Theorem XII (see Note [3.14]).

To avoid any confusion, Jan de Witt gives a strict rule (certa methodos). Apparently he assumes that (*) has been manipulated in such a way that the term v^2 has coefficient $T (\neq 1)$. His instructions are now: first divide i^2v^2/a^2 by the term with v^2 as it occurs in (*): in general this is Tv^2. This gives $i^2/(a^2T)$. Then multiply the whole equation by this result. It is clear that in the result v^2 has coefficient i^2/a^2, which immediately gives the term GB^2. Perhaps unnecessarily he proves this using (*): he divides i^2v^2/a^2 by v^2 (as $T = 1$), which gives i^2/a^2, and then multiplies the whole equation by i^2 and divides the result by a^2.

He seems extremely scrupulous, though we can also see this as a sign of the novelty of the letter-algebra that had just made its appearance.

[3.16] See Note [3.3].

[3.17] The first case is considered from p. [283], line 6 from the bottom, on and has just been dealt with.

[3.18] Jan de Witt again follows the procedure that he sketched on p. [285], line 15 from the bottom, this time with respect to the variable z. This leads him to the first equation on p. [287], which in fact has the form of the equation in Theorem XIII on p. [277], i.e. $y^2 - f^2 = lx^2/g$. Here y must be replaced by z and x by iv/a.

[3.19] The *Regula Universalis* on p. [280] is meant here.

[3.20] See Note [3.3].

[3.21] See the figure on p. [292]. Jan de Witt fails to mention that he restricts himself to the points that lie in a fixed plane with the two given points. He does mention it in Problem I on p. [272], the version of this problem for a parabola. In Problem III, on p. [300], the analogue for an ellipse, he again does not mention it.

[3.22] This is typical for the analytic method.

[3.23] In the Latin text $x = a$ means the absolute value of the difference between x and a; see also Note [2.27].

[3.24] The division by $a^2 - b^2$ is carried out without checking what happens when $a = b$. In that case the locus consists of the abscissa-axis minus the line segment AB. In general Jan de Witt does not consider degenerations.

[3.25] See Note [3.3].

[3.26] This is an allusion to the application of areas. The reference for this is *Liber Primus*, Appendix A.

Both here and on p. [301], the word *figure* (Latin *figura*, Greek εἶδος) has a special meaning, namely the rectangle whose sides are $2a$, the transverse diameter of the hyperbola, respectively of the ellipse, and p, the associated latus rectum. This term was introduced by Apollonius (*Conica* I, 14 and 21).

To determine the points that we call foci (since Kepler), Apollonius proceeds as follows (*Conica* III, 45): a rectangle is applied to the transverse diameter at a vertex, with square excess in the case of a hyperbola and square defect in the case of an ellipse. The area of this applied rectangle is one fourth of the mentioned *figure*: $ap/2$. In the case of a hyperbola, the applied rectangle exceeds the transverse diameter with a square of size $BFHK$ (see Figure 4.6); in the case of an ellipse, a square is missing (see Figure 4.9). If we call this rectangle $AFHL$, then in the case of a hyperbola, a point F arises on the extension of AB if the application originates at A; in the case of an ellipse, a point F arises between A and B. If the application originates at B, then a point G arises in an analogous manner.

The statement is now that the points F and G are the foci that we know. The proof is simple. In this note we give the proof for a hyperbola; we consider the ellipse in Note [3.42].

In Figure 4.6, let $AM = MB = a$ and let the latus rectum be p, then $AF \cdot FH = AF \cdot FB = 2ap/4$ because $FH = FB$. Next let $DM = ME = b$ and $MF = c$, in which case

$$AF \cdot FH = AF \cdot FB = (c+a)(c-a) = \frac{ap}{2}.$$

As p is defined through the proportion $2a{:}2b{=}2b{:}p$ (*Liber Primus*, p. [235]), we have $(c+a)(c-a) = b^2$ and therefore $c^2 = a^2 + b^2$.

It follows that $F(c,0)$ is a focus. The analogous reasoning and conclusion hold for $G(-c,0)$.

From this construction Apollonius deduced in a geometric manner that for a point P on the hyperbola the lines PF and PG make equal angles with the tangent at P (*Conica* III, 48), and that moreover $|PF - PG| = 2a$ (*Conica* III, 51).

In antiquity the foci of an ellipse or of a hyperbola did not have separate names, but were described by Apollonius as

τά εκ της παραβολης γινόμενα σημεια

that is, as *the points that arise through the application.* As was already mentioned in Note [2.28], he does not speak of the focus of a parabola at all.

[3.27] Jan de Witt uses the term *opposite hyperbolas* for the two branches of the hyperbola.

[3.28] The remark that the truth of Corollary I is "completely clear from what we stated before" is justified in lines 3–6 from the bottom of the translation of p. [292] on p. 158.

Indeed, these lines include the deduction from the equation of the hyperbola that the length of the transverse axis is equal to b, which is precisely the constant difference between the distances from the required points to the fixed points A and B. In the analogon of this problem for the ellipse, Jan de Witt does not make any corresponding remark, even though in that case the same holds, mutatis mutandis.

When speaking of "many digressions and by a long concatenation of difficult theorems," Jan de Witt may have had in mind the proofs of Theorems III 51 and 52 by Apollonius, which consist of six preliminary theorems, namely, III 42, 45, 46, 48, 49, 50.

[3.29] See the figure on p. [292] and also p. [293], line 4 from the top.

First CE is drawn perpendicular to the support of the axis; that is, this CE is applied ordinate-wise to the axis.

If, as usual, we denote the semi-transverse diameter ($= HF = HG$) and semi-conjugate diameter by a and b respectively (we abandon the notation of Jan de Witt), then according to p. [293], line 14 from the bottom, we have

$$FB \cdot BG = GA \cdot AF = b^2. \tag{1}$$

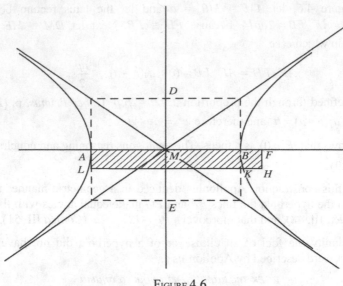

FIGURE 4.6

The point M is now chosen on the extension of FG in such a way that

$$HE : HM = HF : HA, \text{ and therefore } AH \cdot HE = FH \cdot HM,$$

which Jan de Witt as usual announces by writing that the rectangle AHE is equal to the rectangle FHM.

Next Jan de Witt applies Theorem IX, Proposition 10 of *Liber Primus* (p. [196]), which is recalled in Note [3.3] above. In our case this implies that

$$FE \cdot EG : CE^2 = a^2 : b^2 . \text{ Together with (1) this gives}$$

$$FE \cdot EG : CE^2 = HF^2 : GA \cdot AF . \text{ It follows that}$$

$$FE \cdot EG : (FE \cdot EG + CE^2) = HF^2 : (HF^2 + GA \cdot AF). \tag{2}$$

Jan de Witt uses the expression *compositio rationis contraria*, which means "the counterpart of the composition of a ratio."

The concept of composition of a ratio (*compositio rationis*, σύνθεσις λόγου) is defined by Euclid (*El.* V, Def. 14) as the conversion from the ratio $a : b$ to the ratio $(a + b) : b$. This operation uses addition. Jan de Witt uses the counterpart of this, that is, the conversion from $a:b$ to $a:(a+b)$, which does not occur explicitly in the work of Euclid.

In addition to this composition of ratios, Euclid also uses the separation of a ratio (*separatio rationis*, διαίρεσις λόγου, *El.* V, Def. 15). This concept is defined as the conversion from the ratio $(a+b):b$ to the ratio $a:b$. If $a > b$, this can be seen as the conversion from the ratio $a:b$ to $(a-b):b$. This operation uses subtraction.

Fig. i.

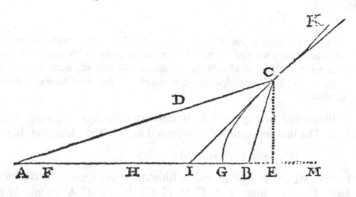

FIGURE 4.7

Another notion is the conversion of a ratio (*conversio rationis*, αναστροφή λόγου, *El.* V, Def. 16). This concept is defined as the conversion from the ratio $(a+b):b$ to the ratio $(a+b):a$. If $a > b$, then this can be seen as the conversion from the ratio $a:b$ to the ratio $a:(a-b)$. Once again the operation uses subtraction.

To sum up, the following conversions are defined:

(i) from $a:b$ to $(a+b):b$ and from $a:b$ to $a:(a+b)$;

(ii) from $a:b$ to $(a-b):b$ and from $a:b$ to $a:(a-b)$.

Euclid shows that a proportion remains valid whenever one of these operations he has introduced is applied to both members.

As an aside, Jan de Witt remarks that

$$HF^2 + GA \cdot AF = HA^2. \tag{3}$$

Let us first make two remarks regarding the text:

(i) Fermat already used the notation *HFq* for HF^2. William Oughtred (1575–1660) introduced the notations *Aq* for A^2, *Ac* for A^3, and combinations of these.

(ii) Jan de Witt writes *HFq ad (HFq + GAF*, id est, ad) *HAq*. The core of the statement is *HFq ad HAq*, that is, $HF^2 : HA^2$. In parentheses Jan de Witt states that *HFq + GAF* is equal to *HAq*. These parentheses are therefore part of the text and not of the mathematical formula. This is why *id est* is also inside the parentheses. See also Note [3.6] concerning the parentheses. We could prove (3) by remarking that $GA = AF + 2HF$ and therefore

$$HF^2 + GA.AF = HF^2 + (AF + 2\,HF).\,AF = (HF + AF)^2 = HA^2.$$

Nevertheless for this conversion Jan de Witt refers, in Marginal Note (3), to *El.* II, 6, which reads:

> If a line segment is cut in half and another segment is adjoined to it, then the rectangle enclosed by the whole line segment with extension and the extension itself together with the square on the half-segment, is equal to the square on the line segment composed of the half-segment plus the extension.

This is illustrated in Figure 4.8. It concerns the line segment *FG*, whose midpoint is *H*. The line segment *AF* is adjoined to *FG*. The following then holds:

$$AG \cdot AF + FH^2 = AH^2 .$$

The proof uses geometric algebra and is illustrated in Figure 4.8. We have *FG* = *HG* and *AA'*= *AF*; the angles at *A, F, H, G, G', H', F', A', K, L,* and *M* are right angles. The equality of the arched areas *HGG'H'*, *FHH' F'*, and *A'F'LK* implies that the rectangle *AGG'A'*, which we will call *R*, has the same area as the edge (*gnomon*) *AFHH' F'LKA'A*, which we will call *T*. We see that

$$\text{area } R + \text{area } F'H'ML \;=\; \text{area } T + \text{area } F'H'ML$$
$$=\; \text{area } AHMK = AH^2 .$$

As *AF* = *AA'*, we have

$$\text{area } R = AA' \cdot AG = AG \cdot AF$$

and therefore $AG \cdot AF + FH^2 = AH^2 .$

If we set *FH* = *HG* = *a* and *AF* = *b*, then we have proved the formula

$$(b + 2a)b + a^2 = (a+b)^2 .$$

Analogously we can prove that

$$HF^2 + FE.EG = HE^2 \quad \text{(Figure 4.7)}. \tag{4}$$

It is worth noting that Jan de Witt does not use the technique of letter-algebra, which was catching on in his days.

Using (2) and (3) we therefore have:

$$FE \cdot EG : (FE \cdot EG + CE^2) = HF^2 : HA^2 .$$

Jan de Witt now applies the rule that *a : b* = *c : d* implies (*a* + *c*) : (*b* + *d*) = *c : d*. This gives

$$(HF^2 + FE \cdot EG) : (HA^2 + FE \cdot EG + CE^2) = HF^2 : HA^2 .$$

Together with (4) this implies

$$HE^2 : (HA^2 + FE \cdot EG + CE^2) = HF^2 : HA^2 .$$

Now *M* was chosen in such a way that

$$HE : HM = HF : HA, \text{ so that } HE^2 \;:\; HM^2 = HF^2 \;:\; HA^2 .$$

These last two lines imply that   ˌ

$$HM^2 = HA^2 + FE \cdot EG + CE^2$$

and $$HM^2 + HF^2 = HA^2 + HF^2 + FE \cdot EG + CE^2,$$

which together with (4) gives

$$HM^2 + HF^2 = HA^2 + HE^2 + CE^2. \tag{5}$$

We augment (in a different situation we would diminish) the left-hand side respectively the right-hand side by $2HM \cdot HF$ respectively $2HA \cdot HE$, which are the same by the choice of M. In our case this leads to

$$(HM + HF)^2 = (HA + HE)^2 + CE^2 = AE^2 + CE^2,$$

so that $$FM^2 = AC^2, \text{ that is: } FM = AC.$$

If we augment (or diminish) the two sides of (5) by $2HM \cdot HG$ and $2HA \cdot HE$ respectively, which are equal to each other because $HG = HF$ and $HM \cdot HF = HA \cdot HE$, then we find

$$GM = BC$$

and therefore $$AC - BC = FM - GM = FG.$$

[3.31] Both figures on p. [295] deal with the case where K lies to the right, respectively to the left, of CI; on p. [291] we mentioned that AD = FG, the transverse axis.

[3.32] $DC = AC - AD$ and $AD = FG = AC - BC$, so $DC = BC$; moreover $\angle DCK$ has supplementary angle $\angle DCI$ and $\angle BCK$ has supplementary angle $\angle BCI$, where $\angle DCI = \angle BCI$.

[3.33] $AL - LB = FG = AD$, $AK = AL + LK$, so $AK - (BL + LK) = AD$, but $BL + LK > BK$ and $BK = KD$, so $AK - KD > AD$, which is impossible.

[3.34] Of course a and b are assumed to be positive; $a < b$ would give a hyperbola and $a = b$ a parabola.

[3.35] We first draw a line through L parallel to AK and then choose a point B on it so that $KL : LB = a : b$.

[3.36] This concerns the last equation on p. [296]. For an analogous reasoning see Notes [3.11] and [3.15].

[3.37] The choice for the ratio $CF : FN$ can be explained as follows: if the equation on the ordinate-wise applied diameters is

$$\frac{x^2}{\alpha^2} + \frac{y^2}{\beta^2} = 1,$$

then the transverse diameter is 2α, the conjugate diameter is 2β, and the latus rectum p defined by $2\alpha : 2\beta = 2\beta : p$ satisfies

$$2a : p = \alpha^2 : \beta^2.$$

as we already saw in [3.3].

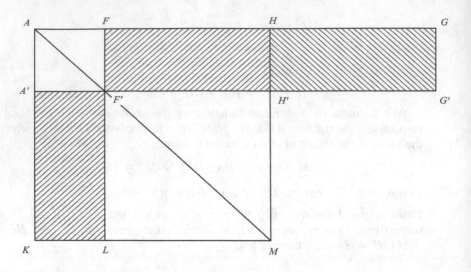

FIGURE 4.8

In our case the equation is

$$\frac{le^2z^2}{ga^2} = \frac{f^2e^2}{a^2} - \frac{e^2v^2}{a^2},$$

where the role of x and y is taken by

$$\frac{ev}{a} = x \text{ and } z = y,$$

so that $$\alpha^2 = \frac{f^2e^2}{a^2}, \ \beta^2 = \frac{ga^2}{le^2}\frac{f^2e^2}{a^2} = \frac{f^2g}{l},$$

and therefore $$2a : p = \alpha^2 : \beta^2 = \beta^2 \frac{f^2e^2}{a^2} : \frac{f^2g}{l} = le^2 : ga^2.$$

If we set $FN = p$, then in the figure on p. [297] (see also p. [298], line 11 from the bottom), we have

$$CF : FN = 2a : p = le^2 : ga^2 = le^2 : ga^2.$$

[3.38] If in the figure on p. [299] we let the transverse diameter FC equal 2α, the conjugate diameter equal 2β, and the latus rectum FN equal p, then it follows from the definition of p, $2\alpha : 2\beta = 2\beta : p$, that $p = 2\alpha$ implies $\alpha^2 = \beta^2$ and therefore $\alpha = \beta$. The equation of the curve on the axes FGC and GH then becomes $x^2 + y^2 = a^2$. As the axes are perpendicular to each other, $x^2 + y^2$ is the square of the distance from a point $D(x,y)$ on the curve to G and is therefore constant, so that we have a circle.

[3.39] The letters A and B both occur twice in Figure I on p. [300]; we will restrict ourselves to the lower ones. Let us note that Jan de Witt fails to mention that he is only interested in a set of coplanar points.

[3.40] If $a = b$, then the locus is the line segment AB.

[3.41] In this formula Jan de Witt mistakenly writes x instead of v. The reference to Theorem XIII is incorrect: this should be Theorem XIV on p. [279].

[3.42] The situation is analogous to that in Note [3.26], where the construction of the foci of a hyperbola by Apollonius is explained. This time we consider the foci of the ellipse; see Figure 4.9. This figure represents the ellipse with symmetry axis AB (= $2a$), conjugate axis DE (= $2b$), and latus rectum p. The rectangle $AFHL$ has been applied to the axis AB at the point A. It has area $2ap/4$, one-fourth of the "Figure," and has square defect $FBHK$ (see Appendix A of *Liber Primus*). If we also set $MF = c$, then

$$AF \cdot FH = AF \cdot FB = (a+c) \cdot (a-c) = \frac{ap}{2}.$$

As $p = 2b^2/a$ (*Liber Primus*, p. [235]), however, we have $a^2 - c^2 = b^2$.

It follows that $F(c,0)$ is a focus. An analogous reasoning and conclusion hold for $G(-c,0)$, which arises from applying a rectangle at B.

From this construction, Apollonius deduces in a geometric manner that for a point P on the ellipse, the lines PF and PG make equal angles with the tangent at P (*Conica* III, 48), and moreover that $PF + PG = 2a$ (*Conica* III, 52).

[3.43] Here Jan de Witt uses *subductio* for subtraction and *divisio* (separation) for the standard term *conversio*, which he uses later on. See also Note [3.30].

[3.44] For the changes that must be made in the proof, see Note [3.45].

[3.45] This corollary is the analogue of Corollary 1 of Problem II on p. [293]. Again a point M is constructed on the transverse axis in such a way that $CA = FM$ and $CB = MG$; in this case, however, the conclusion is that $CA + CB = FG$ (see the figure on p. [302] and Figure 4.10 below).

First CE is drawn perpendicular to the symmetry axis; in other words, CE is applied ordinate-wise on this axis. Then CN is drawn perpendicular to the conjugate axis OP; in other words, CN is applied ordinate-wise on the conjugate axis.

It follows from the remark on p. [301], lines 13 and 14 from the bottom, that

$$FA \cdot AG = GB \cdot BF = HO^2.$$

A point M is now chosen on the transverse axis in such a way that

$$HE : HM = HF : HA \text{ and therefore } AH \cdot HE = FH \cdot HM \qquad (1)$$

Jan de Witt again announces this by stating that the rectangle AHE is equal to the rectangle FHM.

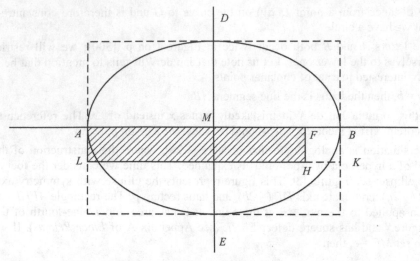

FIGURE 4.9

Let us note that the point M always lies between the points E and H. A point Q is chosen on HG so that $HQ = HE$. It follows from (1) that

$$HE^2 : HM^2 = HF^2 : HA^2 .$$

Jan de Witt now uses the "conversion" of a ratio (*conversio rationis*, see Note [3.30]), which gives

$$HE^2 : (HE^2 - HM^2) = HF^2 : (HF^2 - HA^2) \qquad (2)$$

and therefore $HE^2 : EM.MQ = HF^2 : GA.AF .$ \hspace{1cm} (3)

These last conversions require an explanation. We easily realize that

$$HE^2 - HM^2 = (HE - HM)(HE + HM) = EM.MQ . \qquad (4)$$

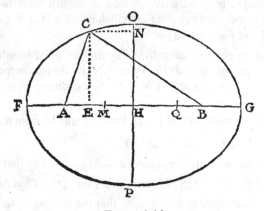

FIGURE 4.10

and analogously

$$HF^2 - HA^2 = GA.AF.$$ (5)

However, as is evident from his marginal notes, Jan de Witt deduces this result using the geometric algebra of Euclid, and for the conversions in (3) he calls upon the *Elements* II, 5; see also Note [3.30].

The result of Euclid states

> If a line segment is cut into equal and unequal parts, then the rectangle enclosed by the unequal parts of the whole, together with the square on the segment between the points of section, is equal to the square on the half.

The proof of this statement is illustrated in Figure 4.11; it concerns line segment *FG*. The point *H* divides it into two equal parts, *FH* and *HG*, while the point *A* divides it into two unequal parts, *FA* and *AG*. The angles at *F*, *A*, *H*, *G*, *G'*, *H'*, *A'*, *F'*, *K*, *L*, and *M* are right angles and *FF'=FA*.

We then have

	area *FH H' F'*	=	area *HGG'H'*
and	area *AH H' A'*	=	area *F'A'LK,*
hence	area *gnomon FAHH'A'LKF'F* =		
	area *AGG'A'* = *GA·AA'* = *GA·AF.*		
As	area *gnomon* + area *A'H'ML* = area *FHMK,*		
it follows that	$GA·AF + HA^2 = HF^2.$		
If we set	$GH = HF = a$ and $AF = b$, then it follows that		
	$(2a - b)\,b + (a - b)^2 = a^2.$		

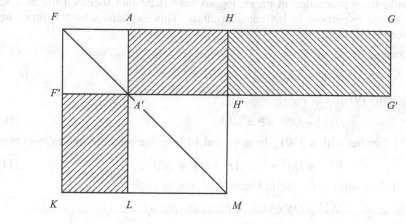

FIGURE 4.11

We apply this theorem to the line segment FG in the figure on p. [302] (= Figure 4.10); again H is the midpoint and A the section point for the unequal parts. Hence in this figure we also have

$$GA.AF + HA^2 = HF^2$$

and therefore $HF^2 - HA^2 = GA.AF$.

Likewise the line segment EQ in the same figure is divided in half by H and in two unequal parts EM and MQ by M. We now have

$$EM.MQ + HM^2 = HE^2$$

and also $HE^2 - HM^2 = EM.MQ$. Finally, it follows from (2) that

$$HE^2 : EM.MQ = HF^2 : GA.AF. \tag{6}$$

We saw on p. [301], lines 13 and 14 from the bottom that $GAF = HO^2$, which transforms (6) into

$$HE^2 : EM.MQ = HF^2 : HO^2. \tag{7}$$

If we see OP as the transverse axis of the ellipse and GF as the conjugate axis, then the characteristic property of an ellipse (see Note [3.5]) implies that

$$CN^2 : ON.NP = HF^2 : HO^2 \tag{8}$$

and, as $CN = HE$, (7) and (8) imply

$$HE^2 : EM.\ MQ = HE^2 : ON.NP,$$

so that

$$EM.MQ = ON.NP. \tag{9}$$

Let us remark that in his treatment of Corollary 1 to Problem II on p. [292], Jan de Witt could have proceeded in an analogous way if he had included the second branch of the hyperbola in his considerations. This explains why the proof he announces is different.

According to (4), we have

$$HM^2 + EM \cdot MQ = HE^2.$$

Consequently (9) gives

$$HM^2 + ON \cdot NP = HE^2. \tag{10}$$

From (5) together with p. [301], lines 13 and 14 from the bottom, now follows that

$$HF^2 = HA^2 + GA \cdot AF = HA^2 + HO^2. \tag{11}$$

But $El.$ II, 5, applied to $ONHP$ in Figure 4.10, implies that

$$HO^2 = ON \cdot NP + NH^2 = ON \cdot NP + CE^2.$$

Hence

$$HF^2 = HA^2 + CE^2 + ON \cdot NP. \tag{12}$$

Equations (10), (11), and (12) imply

$$HM^2 + ON \cdot NP + HF^2 =$$

$$HE^2 + HF^2 = HE^2 + HA^2 + CE^2 + ON \cdot NP,$$

and therefore $HM^2 + HF^2 = HE^2 + HA^2 + CE^2$ (13)

and also $HM^2 + HG^2 = HE^2 + HB^2 + CE^2$.   (14)

If E lies between F and H, then in (13) we should subtract $2HF.HM$ on the left and $2HE.HA$ on the right (by definition of M these terms are equal). The result is then

$$(HF - HM)^2 = (HA - HE)^2 + CE^2,$$

so that $FM^2 = AE^2 + CE^2 = AC^2$, and therefore $FM = AC$.

If E lies between H and G, then in (13) we should add $2HF \cdot HM$ on the left and $2HE.HA$ on the right.

If we apply the analogous operations in (14) with $2HG \cdot HM$ and $2H \cdot .HB$ respectively, then we find

$$GM^2 = EB^2 + CE^2 = CB^2 \text{ and therefore } GM = CB.$$

It is then clear that

$$CA + CB = FG.$$

[3.46] This is the first place where the terms *major axis* and *minor axis* are used.

[3.47] See p. [301], lines 13 and 14 from the bottom.

[3.48] By assumption, $\angle ACI = \angle BCK$ and $\angle LCK = \angle ACI$.

Chapter IV

[4.1]   The plural of "locus" (m.) is usually "loca" (n.) instead of "loci" (m.).

[4.2]   A marginal note states that the sign ℞ means ±. This is the first occurrence of this sign. Van Schooten used it up to 1659 and used the inverse for ∓. Our sign ± was already common in the seventeenth century; see also Cajori I, §210.

[4.3]   See pp. [248] and [249].

[4.4]   The "•"sign, in $y^2 = dx \bullet f^2$ for example, means ±, but Jan de Witt also uses it to mean $y^2 = -dx + f^2$, for example on p. [308], line 2 from the bottom. Obviously $y^2 = -dx - f^2$ is impossible, as the coefficients, in this case d, are positive as they represent line segments. This notation comes from Descartes; see *La Géométrie*, p. 383.

[4.5]   This concerns the substitutions

$$z = y \pm \frac{bx}{a} \pm c, \text{ combined with } v = x \pm c \text{ and}$$

$$v = x \pm \frac{by}{a} \pm c, \text{ combined with } z = y \pm c.$$

Here the coefficients a, b, and c are positive, as they are line segments; the coefficients bx/a and c may be missing. The fraction b/a is imposed by the homogeneity condition, which Descartes has dropped. As is the case for Descartes, this fraction has a trigonometric meaning, as we have already seen in many examples. The first substitution represents a rotation of the x-axis followed by a parallel translation of the x-axis, combined with a parallel translation of what we would call the y-axis. In the second substitution, x and y have been interchanged.

[4.6] It is clear that the sign in front of lx^2/g and the sign between lx^2/g and f^2 determine the answer to "ellipse or hyperbola?." Page [318], line 15 from the bottom explains what is meant by "the difference among the angles."

Indeed, when for an ellipse the conjugate axis of an axis is perpendicular to that axis and the latus rectum p is equal to the transverse diameter $2a$, then the definition of the latus rectum associated with the axes $2a$ and $2b$ (that is, $2a : 2b = 2b : p$) implies that $2a = 2b$, so that we have a circle. See also Note [3.38].

[4.7] It again follows from this that the abscissa x and the ordinate y are seen as line segments. It is worth remarking that Jan de Witt once more emphasizes the meaning of the origin and of the abscissa-axis.

[4.8] Note that only what lies "above" the abscissa-axis counts.

[4.9] The point H is already mentioned on p. [306], line 11 from the bottom, where it is still an arbitrary point on the line BE. It is only fixed on p. [307], line 3 from the top. This process of introducing a point that is only fixed later is often used in the book.

[4.10] See Note [4.8].

[4.11] Jan de Witt follows his usual procedure: first he uses the coefficients a, b, and c that occur in the equations to construct a number of straight lines, and then he proves that the abscissa and ordinate of every point on these lines satisfy the corresponding equation. He used this procedure before, and now applies it in all of Chapter IV. He again works with the terms *determinatio* (the definition of the curve in question) and *demonstratio* (the proof that the abscissa and ordinate of the points on this curve indeed satisfy the equation). We remarked before that in nearly all cases the proof that every point whose abscissa and ordinate satisfy the equation in question lies on the corresponding curve is missing. Finally, note that the figure on p. [307] is not to scale, as we have $AB : BD = a : b$ and $AB : BH = a : b$, while the figure seems to imply that $BD > BH$.

[4.12] From this point on, up to p. [314], line 4 from the bottom, the equations in the first column of case 2 of p. [305] are being considered. These represent parabolas. First curves are constructed using the coefficients that occur in the equations, following a procedure mentioned in the text, the *determinatio*. Nine cases are distinguished, of which most are divided into subcases. This description continues until p. [311], line 5 from the top; the summary contains an overview of these.

To each *determinatio* corresponds a *demonstratio*; these can be found for each of the nine cases from p. [311], line 6 from the top, to p. [314], line 4 from the bottom. The second column on p. [305] is considered more succinctly from p. [314], line 3 from the bottom, to p. [318], line 9 from the top; these cases simply correspond to the interchange of the abscissa and the ordinate.

[4.13] In this chapter Jan de Witt illustrates his argument by means of a number of figures that in the first instance appear to be very complicated, partly due to the fact that the same letter is used for many different points. See for example the figure on p. [308] (= Figure 4.12). Yet there is clearly a system to this, which makes it possible to read the figures. Namely, points that are indicated by the same letter have the same significance for the curve in question: each of them is the vertex or center, the extremity of an axis, etc., but for curves in different positions.

In fact, one figure in this chapter represents a great number of situations, printed one on top of the other.

As Jan de Witt remarks, here and elsewhere in the book he always works with a fixed point A, a fixed line AB with variable B on it, and a fixed angle ABE; AB and BE again denote the abscissa x and the ordinate y of the points that occur.

At first the point A is either the vertex of a parabola or the center of a hyperbola or ellipse, while AB is the support of a symmetry axis.

We take Figure 4.12 as an example. In this case we consider parabolas. At first the vertex lies at A, but as we progress, the vertex is moved to the right or to the left. In both cases the vertex in the new position is called F. The vertex is also moved up and down, in which case the resulting vertex is called D. A combination of these translations leads to four new possible positions for the vertex, which is then called L. The axis can also be rotated, either clockwise or counter-clockwise.

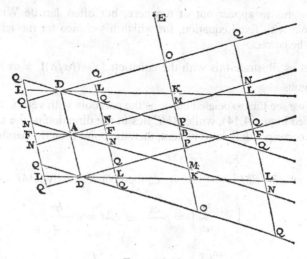

FIGURE 4.12

In both positions the axis is called ANM; here, N is the intersection point with the line through F that runs parallel to BE.

The combination of a translation up or down and a rotation over a positive or negative angle leads to four new positions of the axis, which is then called DQ. If each time we consider the vertex or center of the curve in question in its new position, then from there on we can read the figure without too much trouble, at least if we ignore the letters that are irrelevant.

[4.14] It is only stated at the end of Case IV (on p. [309]) that d is the latus rectum.

[4.15] This second case is divided up into three subcases:

$$\text{(i) } y^2 = dx + f^2, \quad \text{(ii) } y^2 = dx - f^2, \quad \text{(iii) } y^2 = -dx + f^2.$$

The notation
$$y^2 = d\left(x + \frac{f^2}{d}\right)$$

etc. would clarify things, but as was said before, parentheses were not commonly used yet. Figure 4.13 illustrates the situation.

[4.16] The text erroneously states $z^2 = dx f^2$ instead of $z^2 = dx \bullet f^2$.

[4.17] The plural "lines" is used because F can lie both to the right and to the left of A.

[4.18] The lines AM in the "downward" direction are meant here.

[4.19] We now consider the lines with equations

$$y = -\frac{bx}{a} + c \text{ and } y = -\frac{bx}{a} - c,$$

that is, both lines DQP

[4.20] This concerns the case where f^2 does occur.

[4.21] This seems to appear out of nowhere, but often Jan de Witt works towards a parabola with fixed equation, for which his choice for the latus rectum will turn out to be correct.

Let us illustrate this with the equation $[y - (bx/a)]^2 = dx$; the other cases are analogous.

If we see the associated curve as the parabola with vertex A and symmetry axis AM (see Figure 4.14), while BME lies in the direction of the conjugate axis and p is the corresponding latus rectum, then the point E on the parabola satisfies

$$EM^2 = p.AM.$$

If we set $AB : BM : AM = a : b : e,$ then consequently $AM = ex/a$, and

$$\left(y - \frac{bx}{a}\right)^2 = p\frac{ex}{a}, \text{ as } EM = y - \frac{bx}{a}.$$

We also have $\left(y - \dfrac{bx}{a}\right)^2 = dx$, so that $\dfrac{pe}{a} = d,$

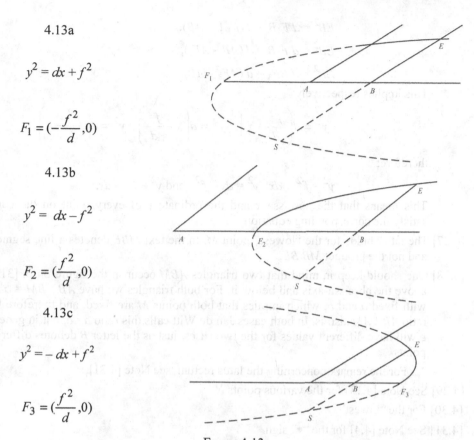

4.13a

$$y^2 = dx + f^2$$

$$F_1 = (-\frac{f^2}{d}, 0)$$

4.13b

$$y^2 = dx - f^2$$

$$F_2 = (\frac{f^2}{d}, 0)$$

4.13c

$$y^2 = -dx + f^2$$

$$F_3 = (\frac{f^2}{d}, 0)$$

FIGURE 4.13

and therefore also $p : d = a : e = AB : AM$.

[4.22] See the end of Note [4.13]. This is where the *demonstratio* begins.

[4.23] To clarify, let us remark the following. The point E is not a fixed point on all parabolas but is the point, which may be different for each parabola, where the parabola meets the line applied ordinate-wise from the point B onto the transverse axis.

[4.24] This concerns the three cases $y^2 = dx + f^2$, $y^2 = dx - f^2$, and $y^2 = f^2 - dx$.

[4.25] In the figure on p. [310] it is the point F that is on the right.

[4.26] Three parabolas have been described, all having a vertex called F: one with the vertex to the left of A, and two with the vertex to the right of A. These vertices are called F whatever their position is; we would speak of F_1, F_2, and F_3 (see the figure in Note [4.15]).

By definition of the constructed parabolas with symmetry axis AB, latus rectum d, and with vertex F_1, F_2, or F_3, we have, in that order,

$$EB^2 = d\, F_1 B = d\,(F_1 A + AB),$$
$$EB^2 = d\, F_2 B = d\,(AB - AF_2),$$
$$EB^2 = d\, BF_3 = d\,(AF_3 - AB).$$

This implies respectively

$$y^2 = d\!\left(\frac{f^2}{d} + x\right);\quad y^2 = d\!\left(x - \frac{f^2}{d}\right);\quad y^2 = d\!\left(\frac{f^2}{d} - x\right);$$

therefore:

$$y^2 = f^2 + dx,\ \ y^2 = dx - f^2,\ \text{and}\ y^2 = f^2 - dx.$$

This means that the abscissa x and the ordinate y of every point on the curve satisfy the corresponding equation.

[4.27] The latter holds for the "lowest" point M. In the text MBE denotes a line segment and not the product $MB.BE$.

[4.28] One should keep in mind that two triangles ABM occur in the figure on p. [313]: above the abscissa-axis and below it. For both triangles we have $AB : BM = a : b$ with fixed a and b, which implies that both points M are fixed, and therefore the ratio $AB : AM$ also is. In both cases Jan de Witt calls this ratio $a : e$, but in general e will have different values for the two cases, just as the letter B denotes different points.

For the remark concerning the latus rectum, see Note [4.21].

[4.29] See Note [4.13] for the various points N.

[4.30] For the "lowest" O.

[4.31] See Note [4.4] for the "•" sign.

[4.32] The text erroneously states $ME = y - (by/a)$ and $MCE = y + (by/a)$.

[4.33] From this point on, up to p. [330], the cases in the first column in 3, p. [305] are considered, as far as they represent hyperbolas. See also Note [4.5]. The second column, which solely concerns hyperbolas, is dealt with from p. [331] to p. [332], line 7 from the bottom.

The cases in the first column that concern ellipses (or circles) are considered from p. [332], line 6 from the bottom to p. [340], line 10 from the top.

[4.34] Thus, in both the "horizontal" and "vertical" case the transverse axis is called FAC.

[4.35] In the equation on the left, the point F is the one on the vertical line through A, which is the support of the symmetry axis of the hyperbola in question. In the equation on the right the point F is the one on the horizontal line through A, which is the support of the symmetry axis of the other hyperbola. Again the same letter is used for the vertex in different positions.

[4.36] For the "lower" D and K.

[4.37] For the "upper" D and K.

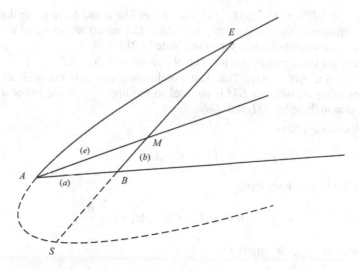

FIGURE 4.14

[4.38] Setting aside the situation where $l = g$, this concerns four cases, where z always satisfies

$$z^2 = \frac{lx^2}{g} + f^2 \text{ or } \frac{lz^2}{g} = x^2 - f^2.$$

After this there is a reference is to the case where f^2 has a plus or minus sign. In that case there are two possibilities for z:

$$z = y \pm \frac{bx}{a}.$$

We will explain these cases some more. The cases where z satisfies

$$z = y \pm \frac{bx}{a} \pm c$$

are analogous and are considered in the text by means of the figure on p. [325], where Jan de Witt needs 23 letters of the alphabet.

Consider Figure 4.15. The following holds:
- Again Jan de Witt often uses a single letter to denote a given point in different positions in the figure. For example, in the figure on p. [325], the letters M and W are associated with the lines AM and EW respectively that are drawn in different directions. This means that extra care is needed to read the text together with the figure. In this note we will differentiate the two points M and W respectively in Figure 4.15 as M_1 and M_2, and W_1 and W_2 respectively. We will then speak of FW_1 and FW_2 where Jan de Witt speaks of FXW and FW. In the first case the first line segment passes through X. We will sometimes also speak of the upper or lower W or M, as this is how the figure must be read.

- $AB = x$, $BE = y$, $AB : BM : AM = a : b : e$. The a and b are those that occur in the equation; this also fixes the ratio $AB : AM$, which we set equal to $a : e$, with two separate possibilities for e (see Note [4.28]).
- There are two possibilities for AM: above AB or below AB.
- $AW \parallel BE$, $EW \parallel AM$. This last condition gives two possibilities for EW, depending on whether EM is parallel to the upper or to the lower AM, which we denote here by AM_1 and AM_2.

In both cases we have

$$EW_1 \text{ (or } EW_2) = EW = AM = \frac{ex}{a}.$$

For the lower W we have

$$AW_2 = AW = M_1E = EB - BM_1 = y - \frac{bx}{a},$$

while the upper W satisfies

$$AW_1 = AW = M_2E = EB + BM_2 = y + \frac{bx}{a}.$$

In his statement of the first case, Jan de Witt speaks of AXW and MBE since the lines pass through X and B.

Finally, points F and C are chosen on AB and AW respectively in such a way that $AC = CF = f$, so that the F on AX satisfies

$$FW_2 = FW = AW_2 + AF = y - \frac{bx}{a} + f = z + f,$$

$$FW_1 = FXW_1 = AXW_1 + AF = y + \frac{bx}{a} + f = z + f,$$

and analogously

$$CW = CW_2 = AW_{21} - AC = y - \frac{bx}{a} - f = z - f,$$

$$CXW = CW_1 = AXW_1 - AC = y + \frac{bx}{a} - f = z - f,$$

Later on we will need the following products:

$$FWC = FW_2 \cdot W_2C = z^2 - f^2$$

and
$$FXWC = FXW_1 \cdot W_1XC = z^2 - f^2.$$

After this follows the *descriptio* of the hyperbolas that Jan de Witt has in mind as being the ones that "satisfy" the equation. Yet he keeps speaking of "the" hyperbola, because in each situation there is one hyperbola that is right.

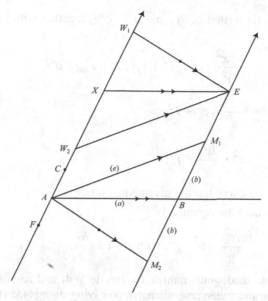

FIGURE 4.15

Here we will restrict ourselves to the hyperbolas that satisfy the equation

$$z^2 = \frac{lx^2}{g} + f^2, \text{ where } z \text{ is defined by } z = y \pm \frac{bx}{a}.$$

For this purpose, a hyperbola is described that goes through E, with center A, transverse diameter on AW and of length $CF = 2f$, and such that AM lies in the conjugate direction. There are two possibilities for this.

The corresponding latus rectum p is determined by the condition that the ratio of the transverse diameter to the latus rectum is $a^2l : e^2g$, so that

$$CF : p = a^2l : e^2g, \text{ that is } 2f : p = a^2l : e^2g.$$

Theorem IX of *Liber Primus* (p. [196]) together with the definition of the latus rectum (see Note [3.3]) implies that

$$EW^2 : FW \cdot WC = p : 2f \tag{i}$$

for the lower W and

$$EW^2 : FXW \cdot WXC = p : 2f \tag{ii}$$

for the upper W.

We now have $EW = \dfrac{ex}{a}$,

and we have already seen that

$$FW_2 \cdot W_2C = FXW_1 \cdot W_1XC - z^2 - f^2$$

The condition on p (that is, $2f : p = a^2l : e^2g$) together with i and ii, then implies that

$$\frac{e^2x^2}{a^2} : (z^2 - f^2) = p : 2f = e^2 g : a^2l,$$

so that

$$\frac{lx^2}{g} = z^2 - f^2$$

and

$$z^2 = \frac{lx^2}{g} + f^2.$$

In both cases the curve has the given equation.

The case where the equation is

$$\frac{lz^2}{g} = x^2 - f^2, \text{ with } z = y \pm \frac{bx}{a},$$

is treated in an analogous manner by Jan de Witt and results in two hyperbolas with center A and transverse diameter NG lying along AM (two possibilities) in such a way that $NA = NG = NA = NG = ef/a$.

The text clearly implies that the hyperbolas defined in the corresponding *descriptio* indeed satisfy the given equation.

[4.39] Jan de Witt speaks of *the hyperbola in question*, but there are in fact two cases, where the transverse diameter lies either on AW or on AM.

[4.40] This concerns the lower M together with the upper W. Jan de Witt conveys this by speaking of MBE and AXE; the points B and X must be "passed through."

[4.41] This concerns the upper M together with the lower W. Jan de Witt simply speaks of ME and AW.

[4.42] The products $FWC = FW.WC$ (lower W) and $FXWC = FXW. WXC$ (upper W) are meant here.

[4.43] This is only a translation over the distance OD of length c, upward or downward, parallel to the direction of BE. The point D takes the place that point A had before (see Note [4.38]).

[4.44] The meaning is clear: in angle DWE the lower point W is the vertex; in $DXWE$ the upper point W is the vertex (we first pass through X). In DOE the upper point O is the vertex; in $DOKE$ the lower point O is the vertex, as we pass through K to E.

[4.45] This concerns four hyperbolas with opening to the right, whose intersection point with BE (the line through B and $\parallel$ to AD) is called E. In all cases the center is called D and the transverse diameter DO. If these points O are named O_1, O_2, O_3, O_4, in downward order, then we easily see that

$$OE \qquad \text{means } O_1E,$$

$$OKE \qquad \text{means } O_2E \text{ (upper } K\text{),}$$

OBE   means O_3E,

OKBE   means O_4E (lower K).

In each of the four cases the following theorem is used:

$$EO^2 : GQ \cdot HO = p : 2a = a^2 g : e^2 l.$$

See also Note [4.38].

[4.46] In this fourth case we still have $v = x \pm h$.

[4.47] See p. [320].

[4.48] The *conjugate* means the mirror image with respect to the line AB. The "casu secundo" three lines down concerns the third subcase (§ 3) on p. [336] of the second case where the locus is a hyperbola (p. [320]).

[4.49] The *descriptio* and *demonstratio* are not announced explicitly, but it is clear that first the hyperbola is constructed, and then the equation $xy = f^2$ is deduced from its geometric properties.

[4.50] $GHBE$ denotes the product $GH.HBE$, in the case where the asymptote GH lies under AB and GHE denotes the product $GH.HE$, for the "upper" GH and HE.

[4.51] If it concerns the I on the left; in IB the I on the right is meant.

[4.52] This again concerns the I on the left, and $IABE$ denotes the product $IAB.BE$.

[4.53] $KGHE$ denotes the product $KGH.HE$, in the case where the asymptote KGH lies above AB and K is the leftmost point on it.

[4.54] $KGHBE$ denotes the product $KGH.HBE$, in the case where the asymptote KGE lies below AB and K is the leftmost point on it.

[4.55] Here Jan de Witt comes back to the formulas in 3 on p. [305].

On pp. [318] to [330], he considered the cases of the formulas with $y^2 = (lx^2/g) \pm f^2$ etc. and showed that these represent hyperbolas. Now he considers the following cases, which concern ellipses:

$$y^2 = -\frac{lx^2}{g} + f^2 , z^2 = -\frac{lx^2}{g} + f^2 ,$$

$$y^2 = -\frac{lv^2}{g} + f^2 , \text{ and } z^2 = -\frac{lv^2}{g} + f^2 .$$

As l and g are assumed positive, f^2 cannot have a minus sign. Again Jan de Witt switches over to $ly^2/g = -x^2 + f^2$ etc.

[4.56] See Note [3.38].

[4.57] This is where the *descriptio* ends. The *demonstratio*, the proof that the coordinates of the points on this curve indeed satisfy the equation in question, follows without being announced.

[4.58] In KBE the lower point K is concerned, in KE the upper point K.

[4.59] Again Theorem XII of *Liber Primus* is applied.

[4.60] Let us illustrate this situation for the case $z = y - (bx/a)$. The case $z = y + (bx/a)$ is analogous.

In the figure on p. [337] (Figure 4.16), the upper M and G and lower N satisfy

$$AB : BM : MA = a : b : e, \quad AF = AC = f \text{ and } FN \parallel CG \parallel BE,$$

so that

$$NA = AG = \frac{ef}{a}, \quad NM = \frac{ef}{a} + \frac{ex}{a}, \quad MG = \frac{ef}{a} - \frac{ex}{a}.$$

We now construct an ellipse with center A and transverse diameter NG in such a way that the latus rectum p satisfies

$$2f : p = e^2 l : a^2 g.$$

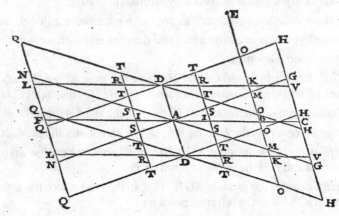

FIGURE 4.16

Because of Theorem XII of *Liber Primus* (see also Note [3.5]), the point E on the ellipse, where EM is applied ordinate-wise on the axis NG, satisfies

$$EM^2 : NM \cdot MG = p : 2f = a^2 g : e^2 l.$$

As

$$EM = y - \frac{bx}{a} = z,$$

we have

$$e^2 l z^2 = a^2 g \left(\frac{ef}{a} + \frac{ex}{a} \right) \left(\frac{ef}{a} - \frac{ex}{a} \right),$$

so that

$$\frac{l z^2}{g} = f^2 - x^2.$$

[4.61] If the lower D is concerned.

[4.62] End of *descriptio*, beginning of *demonstration.*

5

Appendix

5.1. Background of Pappus's problem

The first mention of the problem that would later become known as Pappus's problem can be found in the introduction to the *Conica* (I, 1) by Apollonius. There he writes that Euclid had insufficiently worked out the problem of the locus in the case of three or four lines and that the results of Euclid were not even that good. Apollonius refers to the work of Euclid on conics, which is now lost. He attributes the shortcomings of Euclid to the unavailability of two relevant fundamental theorems that Apollonius himself found and proved. We will see further on which theorems these are.

As we saw in the introduction, the problem meant here, of the locus of three or four lines, can be put into the following simple form.

Let three or four lines $l_1, l_2, l_3(l_4)$ be given in the plane. We are looking for the set of points P whose distances $d_1, d_2, d_3(d_4)$ from respectively $l_1, l_2, l_3(l_4)$ satisfy

(i) $d_1 d_2 : d_3^2$ = constant   (in the case of three lines),

(ii) $d_1 d_2 : d_3 d_4$ = constant   (in the case of four lines).

These distances can be taken perpendicularly, but also in a direction that is determined separately for each line l_i. It is clear that this choice does not change the nature of the problem.

The situation for four lines is illustrated in Figure 5.1. Let us already mention that the loci in question will prove to be conics.

An obvious question is "How did anybody come up with such a problem?"

We can obtain a partial answer by restricting ourselves to the case where the locus is an ellipse; we will use our modern notation. In Figure 5.2, AB is an axis of the ellipse E, of length $2a$; t_1 and t_2 are tangents to the ellipse at A and B, P is an arbitrary point on it, and PP_1 is parallel to t_1 and t_2.

A.W. Grootendorst et al. (eds.), *Jan de Witt's Elementa Curvarum Linearum*,
Sources and Studies in the History of Mathematics and Physical Sciences,
DOI 10.1007/978-0-85729-142-4_5, © Springer-Verlag London Limited 2010

Already in ancient times the following was known to characterize an ellipse:

$$PP_1^2 : AP_1.P_1B = \text{constant}.$$

The size of this constant is $b^2 : a^2$, where $2a$ is the length of the transverse axis AB and $2b$ is the length of the conjugate axis. Jan de Witt also deduced this property from his definition of an ellipse and used it as characterizing property for the ellipse (*Liber Primus*, Th. XII, Prop. 13). If we draw the line $A_1B_1 \parallel AB$ through P, then we also have

$$PP_1^2 : PA_1.PB_1 = \text{constant},$$

so that the points on the ellipse E are the solutions of the three-line problem for these t_1, t_2, and AB, with distances taken in the direction of AB and of t_1.

We can generalize this situation; see Figure 5.3. Here AB is an arbitrary chord of the ellipse E with center M; t_1 and t_2 are the tangents at A and B; their intersection point is T. We now draw a line $A_1B_1 \parallel AB$ through an arbitrary point P of the ellipse, and PP_1 parallel to the axis through T. We can then prove that

$$PP_1^2 : PA_1.PB_1 = \text{constant}.$$

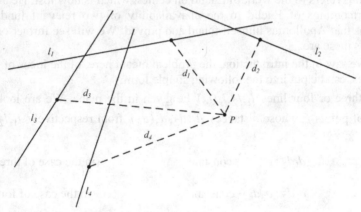

FIGURE 5.1

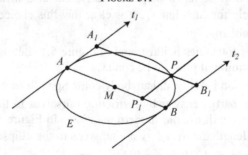

FIGURE 5.2

We can generalize this situation; see Figure 5.3. Here AB is an arbitrary chord of the ellipse E with center M; t_1 and t_2 are the tangents at A and B; their intersection point is T. We now draw a line $A_1B_1 \parallel AB$ through an arbitrary point P of the ellipse, and PP_1 parallel to the axis through T. We can then prove that

$$PP_1^2 : PA_1.PB_1 = \text{constant.}$$

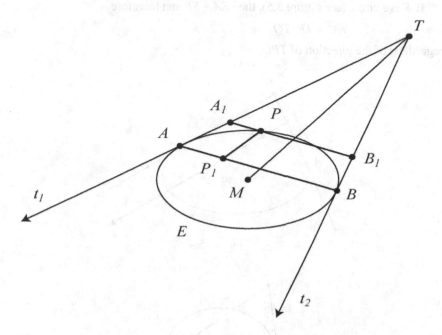

FIGURE 5.3

The ellipse E is therefore the solution of the three-line problem for these t_1, t_2, and AB, with distances taken in the direction of AB and of TM.

Conversely it is not difficult to show that for three given lines and their associated directions the solution of the three-line problem is an ellipse or a hyperbola (Knorr [43], p. 122).

The proofs Apollonius gave for this as well as possible contributions of Euclid have not been passed down, so that we depend on reconstructions. The most important of these were given by Heath [30], Knorr [43], and Zeuthen [70, 71].

We can find the two relevant fundamental theorems needed here, which we mentioned above, in the *Conica* of Apollonius as Theorems III, 16 and III, 17. We recall them here because they are not commonly known, though they are generalizations of two universally known theorems for circles. They amount to the following.

First case (Conica III, 16; see Figure 5.4)

If we draw the tangents *SA* and *SB* from a point *S* to the conic *K*, choose a point *T* on *SA*, and draw a line parallel to *SB* through *T*, which meets the conic at *P* and *Q*, then

$$TA^2 : TP \cdot TQ = SA^2 : SB^2$$

If *K* is a circle (see Figure 5.5), then *SA = SB* and therefore

$$TA^2 = TP \cdot TQ$$

regardless of the direction of *TPQ*.

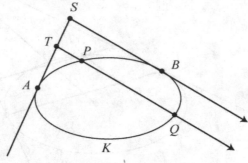

FIGURE 5.4

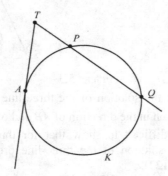

FIGURE 5.5

Second case (Conica III, 17; see Figure 5.6).

If we draw the tangents *SA* and *SB* from a point *S* to the conic *K*, and draw the chords *PTQ* and *RTU* parallel to *SA* and *SB* respectively through a point *T* inside or outside *K*, then

$$PT \cdot TQ : RT \cdot TU = SA^2 : SB^2.$$

If K is a circle (see Figure 5.7), then again $SA = SB$ and therefore $PT \cdot TQ = RT \cdot TU$

Heath and Knorr reconstructed the solution to the three-line problem in different ways (Heath [30], p. 122, Knorr [43], pp. 121–122), but Knorr does agree with the solution to the four-line problem that Heath deduced from his reconstruction (Heath [30], pp. 123–125).

Heath first shows that any conic is a solution for the four-line problem for an inscribed quadrilateral. He does this by applying the solution to the three-line problem four times.

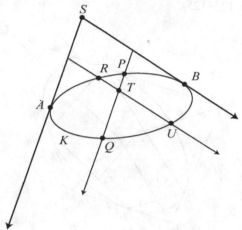

FIGURE 5.6

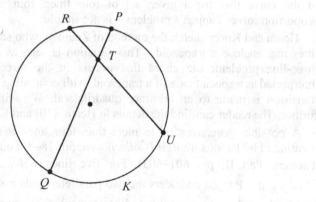

FIGURE 5.7

Let us consider Figure 5.8. In this example the quadrilateral $A_1 A_2 A_3 A_4$ is inscribed in the ellipse E with center M; the tangents at A_1, A_2, A_3, A_4 meet at

T_1, T_2, T_3, T_4, and P is an arbitrary point on E. We draw a line through P that is parallel to T_1M, and meets A_1A_2 at P_1. Analogously, we determine P_2 on A_2A_3, P_3 on A_3A_4, and P_4 on A_4A_1; that is, PP_2 is parallel to T_2M, and so on. By applying the result of the three-line problem four times, to the configurations $T_1A_1A_2, T_2A_2A_3, T_3A_3A_4$, and $T_4A_4A_1$, each time with respect to the point P on E, we find, after some computation:

$$PP_1 \cdot PP_3 : PP_2 \cdot PP_4 = \text{constant};$$

see Heath ([30], pp. 123–125).

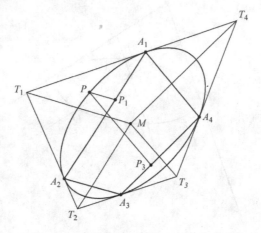

FIGURE 5.8

The converse of this problem, the true Pappus problem – that is, the construction of the curve that for a given set of four lines, four directions and a given proportion solves Pappus's problem – is not simple.

Heath and Knorr sketch the method of Zeuthen, who starts out with four given lines that enclose a trapezoid. This situation is seen as a generalization of the three-line problem; the chord that occurs in the three-line problem is then interpreted as a special case of a trapezoid, with coinciding parallel sides. Then the transition is made to an arbitrary quadrilateral. We will not go into this any further. The reader can find the details in Heath [30] and Knorr [43].

A possible generalization to more than four lines is obvious and is already mentioned by Pappus himself (*Collectio* vii, pp. 38–40; ed. Hultsch p. 680, and I. Thomas, Part II, pp. 601–603). For five lines $l_1, l_2, ..., l_5$, the distances are $d_1, d_2, ..., d_5$. Pappus considers the two parallelepipeds enclosed by d_1, d_2, d_3, and by d_4, d_5, and a randomly chosen line segment a. The condition on P is now that the ratio $d_1d_2d_3 : d_4d_5a$ of the volumes is constant. For six lines the distances are $d_1, d_2, ..., d_6$. The condition is then that the ratio $d_1d_2d_3 : d_4d_5d_6$ is constant. When we have more than six lines, there is a problem that Pappus clearly recognizes. We are faced with products of four or more lines, which do not have any significance in the geometric algebra of the Greeks, as they do not represent any quantity. Nevertheless, Pappus says: "Some modern writers have agreed to speak of them even if what they say is unclear." By this he may have in mind Heron (3rd century AD) and his formula for the area of a triangle, which we know as Heron's formula, in which the product of four line segments occurs.

Pappus does not go further than eight line segments and to circumvent the problem of the ratio of products of four line segments he does not speak of the ratio $d_1 d_2 d_3 d_4 : d_5 d_6 d_7 d_8$, but of the ratio that is "composed" of the ratios $d_1 : d_5, d_2 : d_6, d_3 : d_7$, and $d_4 : d_8$. This requires an explanation, but in short it corresponds to the multiplication of real numbers. The notion of "composition" of ratios first occurs in Euclid's *Elements*, VI, 23. He proves that the areas of two parallelograms with a common angle are in the same ratio as the products of the sides a and b to the product of the sides c and d enclosing the angle, that is, as ab to cd. He states this as follows:

> The areas of parallelograms with a common angle have a ratio that is composed of the ratios of the corresponding sides.

If we denote the ratio of similar quantities x and y by x/y and the operation "composition" by the symbol *, then Euclid in fact says that

$$\frac{ab}{cd} = \frac{a}{c} * \frac{b}{d}.$$

In his proof of this theorem, Euclid tacitly assumes this notion of composed ratios to be known. Dijksterhuis ([18], Part II, p. 83) gives a reconstruction of its definition. From this we have the following.

If $a_1, a_2, ..., a_n$ is a series of similar quantities, then the ratio a_1/a_n is said to be composed of the ratios:

$$\frac{a_1}{a_2}, \frac{a_2}{a_3}, ..., \frac{a_{n-1}}{a_n},$$

that is:

$$\frac{a_1}{a_n} = \frac{a_1}{a_2} * \frac{a_2}{a_3} * ... * \frac{a_{n-1}}{a_n}.$$

We easily deduce from this definition that

$$\frac{abc}{xyz} = \frac{a}{x} * \frac{b}{y} * \frac{c}{z}.$$

Indeed, if we consider the series abc, xbc, xyc, xyz, then according to the definition of composition of ratios:

$$\frac{abc}{xyz} = \frac{abc}{xbc} * \frac{xbc}{xyc} * \frac{xyc}{xyz} = \frac{a}{x} * \frac{b}{y} * \frac{c}{z}.$$

In this same manner we can see the ratio of two products of more than three line segments as composed of the ratios of two line segments at a time. Nevertheless, the problem of the significance of such a product of more than three line segments remains.

Pappus (*Collectio*, vii, 928) proved geometrically that

$$\frac{ab}{cd} = \frac{a}{c} * \frac{b}{d}.$$

As we already saw in the introduction, Descartes would later remove the obstacles the Greeks encountered in their interpretation of products of more than three line segments, by introducing a unit length 1. For the details we refer the reader to the Introduction, Section 1.5.

5.2. Descartes's dissertation on plane curves

In the antiquity, as we mentioned before, plane curves were divided over three classes: plane loci (*loci plani*); straight lines and circles; solid loci (*loci solidi*), conics other than the circle; and finally *loci lineares*, which include all other plane curves, such as the cissoid, conchoid, quadratrix and spiral. Descartes did not want to call these last curves mechanical as all types of instrument were needed to draw them. Yet this also holds for the line and the circle, for which one needs a ruler and a compass.

He proposes to allow curves in the geometry according to the following new axiom:

> Two or more curves can be moved in such a way one through the other that other curves arise through their intersection point. In other words, the curves that are admissible in the geometry are those that can be described through a continuous movement or through successive continuous movements, where each movement is completely determined by that which precedes it. For clarification Descartes gives the example of a curve that arises through the movement of a number of rods where each rod causes the following one to move.

Figure 5.9. illustrates what this means. The ruler *YX* rotates around the point *Y*, *YB* has a fixed length. The rod *BC* is perpendicular to *YX*, and when the ruler *YX* rotates around *Y*, *BC* moves rod *CD* that lies perpendicular to *YZ*. In turn *CD* causes rod *DE* that lies perpendicular to *YX* to slide, which in turn makes *DE* cause rod *EF* (perpendicular to *YZ*) to move, and so on. The intersection point of the last rod, *GH*, with the rotating ruler *YX* then describes the curve that Descartes means: it arises through successive continuous movements.

Descartes merely notes that the locus of *B* is a circle and that the loci of *D*, *F*, and *H* become successively more "complex," without deducing their equations. If one sets *YB* = *a*, *YG* = *x*, and *GH* = *y*, then after some computation one finds the equation of the locus of *H*: $x^{12} = a^2(x^2 + y^2)^5$. We will not go further into this example, but refer for it to *Géométrie*, pp. 317 and 318, and to [16] and [53].

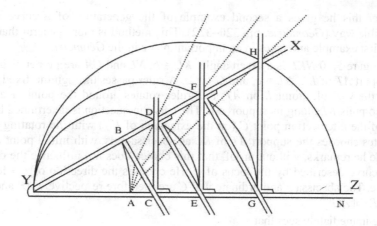

FIGURE 5.9

Even though Descartes does not determine any equations here, he does make the following remark, using a sentence that can be seen as one of the most important in all of mathematical literature:

> *Ie ne sçache rien de meilleur que de dire que tous les poins, de celles qu'on peut nommer Geometriques, c'est .a dire qui tombent sous quelque mesure précise & exacte, ont necessairement quelque rapport a tous les poins d'une ligne droite, qui peut estre exprimé par quelque equation, en tous par une mesme.*

Translated:

> I cannot describe this better than by saying that all points on these [curves] that may be called geometric, that is, which in some way allow for a precise determination, necessarily have a certain relation to the points on a straight line, a relation that can be described through an equation that is the same for all points.

This concerns the relation between the abscissa and the ordinate of any point on the curve in question. Though Descartes does not say it explicitly, what follows implies that the relation must be an algebraic equation. Therefore, for him admissible curves are algebraic curves. Curves that are not included in this were called mechanical curves. Leibniz would later make the distinction between algebraic and transcendental.

Right after this Descartes introduces a classification of geometric curves in what he calls *genres*; we will call these classes.

The first class is made up of the curves that are represented by an equation of degree 1 or 2; that is, lines, circles, parabolas, hyperbolas, and ellipses. The second is made up of the curves with an equation of degree 3 or 4. In general each class consists of the curves with an equation of some odd degree or the consecutive even degree. Descartes is inspired by that fact that an equation of degree 4 can be solved using an equation of degree 3, and by his incorrect conjecture that an equation of degree 6 can be solved using an equation of degree 5, etc.

After this he gives a second example of the generation of a curve in an admissible way (*Géométrie*, pp. 320–322). This method is more general than that of his first example and takes in an important place in the *Géométrie*.

In Figure 5.10 *NKL* is a given angle, $KL = b$, *NL* and *GA* are perpendicular to the support *AT* of *KL*, $NL = c$, and $GA = a$. A ruler passes through the fixed point *G* and the variable point *L* on *AT*; this ruler rotates around the point *G* and in doing so pulls *KL* along its support *AT*. The curve in question is determined by the locus of the intersection point *C* (on the extension of *KN*) with the rotating ruler. Descartes chooses the support *AT* of *KL* as abscissa-axis, with initial point *A*. As an aside he remarks, without proof, that this choice does not influence the degree of the curve described by the locus of *C*. He chooses the direction of *AG* for the ordinate. The abscissa *x* and ordinate *y* of *C* are therefore respectively *AB* and *BC*, where $BC \parallel AG$.

One immediately sees that

$$CB : BK = NL : LK = c : b,$$

hence

$$BK = \frac{by}{c},$$

so that

$$BL = BK - LK = \frac{by}{c} - b$$

and

$$AL = AB + BL = x + \frac{by}{c} - b.$$

We also have

$$CB : BL = GA : AL,$$

hence

$$y : \left(\frac{by}{c} - b \right) = a : \left(x + \frac{by}{c} - b \right)$$

and therefore

$$y^2 = cy - \frac{cxy}{b} + ay - ac.$$

This implies that the curve in question belongs to the first class. Descartes moreover states without further explanation that it concerns a hyperbola. Let us remark that Jan de Witt in principle also used this method to generate the two branches of a hyperbola and deduce a large number of properties from that, in a purely geometric manner (*Liber Primus*, pp. [178]–[182]).

If the moving curve is a half circle with center *L* and radius *R*, then one obtains a curve with equation

$$x^2 y^2 + y^2 (a - y)^2 = R^2 (a - y)^2.$$

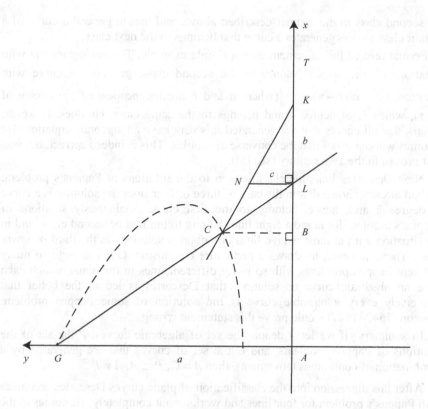

FIGURE 5.10

If one chooses $a - y$ as new ordinate – that is, inverses the direction of the ordinate-axis and takes G as new origin – and then interchanges the abscissa and ordinate, then one obtains the equation of the conchoid in the known form

$$(x-a)^2 (x^2 + y^2) = R^2 x^2.$$

Descartes, however, states without further explanation that this choice of the moving curve leads to a conchoid.

If, instead of the angle NKL one chooses to move a parabola,[1] then a degree 3 curve arises that is known as the parabola or trident of Descartes. It plays an important role in Pappus's problem for five lines and in the solution of equations of degree 5 or 6. We will come back to this curve in Section 5.3 of this appendix.

Descartes now draws a number of conclusions that are not all correct. First he remarks that computations show that a curve of the first class generates a curve of

---

[1] Note by the editors: the (rotating) ruler is attached to the midpoint of the circle of radius R, which moves alongside the ordinate-axis. The curve consists of the points of intersection of the circle and ruler.

the second class in the manner described above, and that in general a curve of a certain class always generates a curve that belongs to the next class.

Fermat refuted this statement with a simple example. The cubic parabola with equation $y^3 = x$, which belongs to the second class, generates a curve with equation $(y^3 - a)(b - y) = xy$ (where a and b are independent of the choice of ruler), which is of degree 4 and belongs to the same class. He does, however, remark that all curves that are generated this way have an algebraic equation. He assumes without proof that the converse also holds. This is indeed correct, but was first proven in the 19th century (see [5]).

Next Descartes links his classification to the solutions of Pappus's problem. He had already shown that in the case of three or four lines the solution is a curve of degree 2 and, hence, belongs to the first class. Analogously solutions of Pappus's problem for at most eight lines belong to the first or second class, and in the situation with at most twelve lines one obtains solutions in the third or lower class. Then, however, he draws a premature conclusion. One can state so many different Pappus problems with so many different lines in the plane, which each have an algebraic curve as solution, that Descartes is led to the belief that conversely every admissible curve is the solution of some Pappus problem. Newton (1642–1747) would prove this statement wrong.

In summary: if we let A denote the set of algebraic curves, P the set of the solutions of Pappus problems, and C the set of curves that are generated by a combination of continuous movements, then $A = C$, $P \subset A$, $A \neq P$.

After this digression into the classification of plane curves Descartes continues with Pappus's problem for four lines and works it out completely. He comes to the conclusion that in this case the solutions are conics, which he describes in detail. The computations that lead to his conclusions have been summed up in the Introduction. We will continue here with Descartes's treatment of Pappus's problem for five lines, which conclude his treatment of Pappus's problem.

5.3. Descartes's treatment of Pappus's problem for five lines

The first case Descartes considers is that where all five lines are parallel. He states without proof that in this case the required points lie on one or more straight lines. We directly see that these lines depend on the solutions of a degree 3 equation with real coefficients.

Next he considers the situation where four lines l_1, l_2, l_3, l_4 are equidistant from each other, with common distance a, while the fifth line l_5 lies perpendicular to the others; see Figure 5.11. Of the possibilities for the position of the distances d_i in the formula, Descartes chooses

$$d_1 d_2 d_4 = a d_3 d_5.$$

Together with the properties of the abscissa x and ordinate y of C that were fixed prior to this, this gives

$$(2a - y)(a - y)(a + y) = axy.$$

The equation of the locus can therefore be written as

$$y^3 - 2ay^2 - a^2 y + 2a^3 = axy.$$

This locus will turn out to be the parabola, or trident, of Descartes that was mentioned before.

To show this Descartes considers the situation in Figure 5.2 and replaces the angle NKL by a convex parabola with axis along AN, vertex at K, and latus rectum a (see Figure 5.12).

The point L on AN lies inside the parabola at a fixed distance a from K. The ruler through G ($GA = 2a$) and L again rotates around G, pulling L along AN. The point L draws the parabola along. Again we are interested in the path of the intersection point C of the support of GL with the moving parabola. At all times CB and CM are perpendicular to respectively AN and AG. In this situation we have

$$GM : CM = CB : LB,$$

so that

$$(2a - y) : x = y : LB,$$

hence

$$LB = \frac{xy}{2a - y}.$$

We also have

$$BK = LK - LB,$$

hence

$$BK = a - \frac{xy}{2a - y}.$$

From the definition of the parabola we then have

$$CB^2 = aBK,$$

and therefore

$$y^2 = a\left(a - \frac{xy}{2a - y}\right); \text{ that is: } y^3 - 2ay^2 - a^2 y + 2a^3 = axy.$$

The equation of this curve is therefore the same as that of the required locus, called the trident. It is easy to see that in general, when $LK = a$, $AG = b$, and the latus rectum of the parabola is p, the constructed curve satisfies

$$y^3 - by^2 - pay + pab = pxy.$$

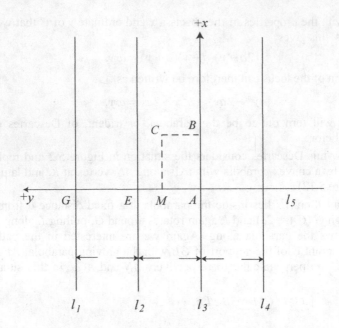

FIGURE 5.11

Finally Descartes remarks that C can also be chosen on the right half of the parabola, and also on the associated convex parabola. He does not give a drawing of the trident he introduced; we have added one here for the case $a = 2$ (Figure 5.13).

Next, the situation where the lines l_1, l_2, l_3, l_4 are not equidistant or perpendicular to l_5, and the lines drawn from C are not perpendicular to the corresponding l_i comes up briefly. This, he says, leads to a solution of an entirely different nature, as does the case in which no two lines are parallel.

It speaks for itself that Descartes also discusses the case where $d_1 d_2 d_3 = a d_4 d_5$ (see Figure 5.13). Here too the solution curve is of an entirely different nature.

Descartes does not give its equation but describes it in a convoluted manner. Rabuel later determined the following equation:

$$axy - xy^2 + 2a^2 x = a^2 y - ay^2$$

(see [16] and [53]).

For a more detailed treatment of Pappus's problem for fives lines and the link to the Cartesian parabola, the reader is referred to the captivating article by Bos ([5]), which also treats the origin of Descartes's method of generating curves by combined continuous movements.

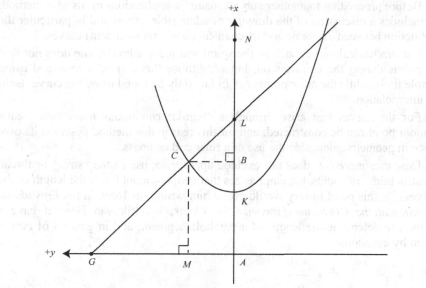

FIGURE 5.12

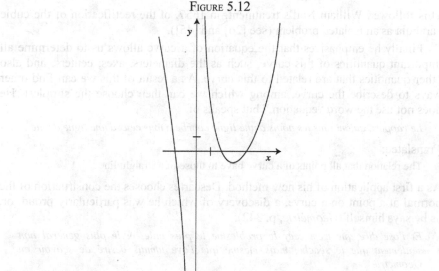

FIGURE 5.13

With this last case of the five-line problem Descartes concludes his treatment of Pappus's problem, because, he says, it was not his intention to treat the problem exhaustively and he considers it sufficient to have given a method to determine an infinite number of points on the solution curve.

Before proceeding to another subject, namely, applications of his new method, he includes a discussion of the drawing of admissible curves and in particular the distinction between algebraic and mechanical (later: transcendent) curves.

For mechanical curves such as the spiral and the quadratrix, one does not find all points during the construction, but only those that can be determined using simple tools, while the other points can in fact only be found using the curve itself, by interpolation.

For the curves that arise through a "regular continuous movement," each random point can be constructed, and for this reason this method deserves its own place in geometry, alongside the use of a ruler and compass.

Descartes moreover does not exclude appropriate use of the "string or thread constructions" of conics, but emphasizes that people cannot know the length of the curves. On this point history would prove him wrong. In 1659, in his *Epistula de Transmutatione Curvarum Linearum in Rectas*, Hendrik van Heuraet gave a method to determine the length of a parabola segment, and in general of curves given by equations

$$y^n = \frac{x^{n+1}}{a}.$$

This followed William Neil's treatment, in 1657, of the rectification of the cubic parabola as an isolated problem (see [26] and [45]).

Finally he emphasizes that the equation of a curve allows us to determine all important quantities of this curve, such as the diameters, axes, centers, and also other quantities that are related to this curve. As a result of this we can find other ways to describe the curve, among which we can then choose the simplest. He does not use the word "equation," but speaks of

Le rapport, qu'ont tous les poins d'une ligne courbe a tous ceux d'une ligne droite.

Translated:

The relation that all points of a curve have to those of a straight line

As a first application of his new method, Descartes chooses the construction of the normal at a point on a curve, a discovery of which he was particularly proud, or, as he says himself (*Géométrie*, p. 342):

Et i'ose dire que c'est cecy le problesme le plus utile, & le plus general non seulement que ie sçache, mais mesme que i'aye iamais desiré de sçavoir en Geometrie.

Translated:

And I dare say that this is not only the most useful and most general problem that I know, but even that I have ever wanted to understand in geometry.

With this we conclude the treatment of Pappus's problem by Descartes. For the application of the parabola of Descartes to the solution of the equation of degree six, which was the main purpose of Descartes, we refer the reader to Bos [6] and Scott [53].

Bibliography

[1] APOLLONIUS, *Apollonii Pergaei quae Graece exstant cum commentaries antiquis edidit et Latine interpretatus est I.L. Heiberg*, Vols. I & II. Stuttgart, Teubner, 1974.

[2] ARCHIMEDES, *Opera Omnia cum Commentariis Eutocii edidit I.L. Heiberg corrigenda adiecit E. S. Stamatis*. Stuttgart, Teubner, 1972.

[3] BOS, H.J.M., On the representation of curves in Descartes's Géométrie. *Archive for History of Exact Sciences* **24** (1981), pp. 295–338.

[4] BOS, H.J.M., *Descartes en het Begin van de Analytische Meetkunde*. CWI syllabus **25**, pp. 79–97. Amsterdam, Centrum Wiskunde & Informatica, 1989.

[5] BOS, H.J.M., Descartes, Pappus's Problem and the Cartesian Parabola, a Conjecture, *Festschrift for D.T. Whiteside*, ed. P. Harman and A. Shapiro. Cambridge, C.U.P.,1992

[6] BOS, H.J.M., The Structure of Descartes's Géométrie. *Lectures in the History of Mathematics* **7**, pp. 37–57. Providence, Rhode Island, American Mathematical Society, 1997.

[7] BOS, H.J.M., *Redefining Geometrical Exactness*. New York etc., Springer-Verlag, 2001.

[8] BOYER, C.B., *History of Analytic Geometry*. New York, Scripta Mathematica, 1956.

[9] BUSARD, H.L.L., *Nicole Oresme, Quaestiones super Geometriam Euclidis*. Leiden, Brill, 1961.

[10] COOLIDGE, J.L., *The Mathematics of Great Amateurs*. New York, Dover Publications, 1963.

[11] COOLIDGE, J.L., *A History of the Conic Sections and Quadratic Surfaces*. New York, Dover Publications (reprint), 1968.

[12] COOLIDGE, J.L., *A History of Geometrical Methods*. New York, Dover Publications (reprint), 1963.

[13] CUOMO, S., *Pappus of Alexandria and the Mathematics of Late Antiquity*. Cambridge, University Press, 2000.

[14] DESCARTES, RENÉ, *Oeuvres de Descartes*, edited by Ch. Adams and P. Tannery, Vol. VI, 1. Paris, Vrin/CNRS, 1964–1976.

[15] DESCARTES, RENÉ, *Œuvres Philosophiques*, edited by F. Alquié. Paris, Garnier Frères, 1963, 1967, 1973.

[16] DESCARTES, RENÉ, *The Geometry of René Descartes*, translated from the French and Latin by David Eugene Smith and Maria L. Latham. New York, Dover Publications, 1954.

[17] DESCARTES, RENÉ, *Geometrie,* German, edited by Ludwig Schlesinger. Darmstadt, Wissenschaftliche Buchgesellschaft, 1969.

A.W. Grootendorst et al. (eds.), *Jan de Witt's Elementa Curvarum Linearum,*
Sources and Studies in the History of Mathematics and Physical Sciences,
DOI 10.1007/978-0-85729-142-4, © Springer-Verlag London Limited 2010

[18] DIJKSTERHUIS, E.J., *De Elementen van Euclides I&II.* Groningen, Noordhoff, 1930.

[19] DIJKSTERHUIS, E.J., *Archimedes I&II.* Groningen, Noordhoff, 1938. Translated into English by C. Dikshoorn. Copenhagen, Ejnaar Munksgaard, 1956.

[20] EUCLIDES, *Euclidis Elementa*, post I.L. Heiberg edidit E. S. Stamatis, Vols. I–V. Leipzig, Teubner, 1969–1977.

[21] FERMAT, PIERRE DE, *Œuvres.* Paris, Gauthier Villars, 1891–1912.

[22] FLADT, K., *Geschichte und Theorie der Kegelschnitte und der Flächen zweiten Grades.* Stuttgart, Ernst Klett Verlag, 1963.

[23] FRUIN, R., *Brieven van Johan de Witt*, edited by R. Fruin and published by G.W. Kernkamp and N. Japikse Vol. IV. Amsterdam, Johannes Müller, 1913.

[24] GEER, P. VAN, Johan de Witt als wiskundige. *Nieuw Archief voor Wiskunde* (2), XI, (1915), pp. 98–126.

[25] GILLESPIE, C.C. (ed.), *Dictionary of Scientific Biography* (VIII Vols.). New York, Charles Scribner's Sons, 1981.

[26] GROOTENDORST, A.W., *Grepen uit de Geschiedenis van de Wiskunde.* Delft, D.U.P., 1988.

[27] GROOTENDORST, A.W., De Meetkundige Algebra bij Euclides. *CWI Syllabus* **28**, pp. 1–26. Amsterdam, Centrum Wiskunde & Informatica, 1991.

[28] GROOTENDORST, A.W., De Kegelsneden bij Jan de Witt. *Nieuw Archief voor Wiskunde* (4), XVII, (1999), pp. 409–425.

[29] HEATH, SIR THOMAS L., *The Thirteen Books of Euclid's Elements*, translated from the text of Heiberg, with Introduction and Commentary by Sir Thomas L. Heath, Vols. I–III. New York, Dover Publications, 1956.

[30] HEATH, SIR THOMAS L., *Apollonius of Perga, Treatise on Conic Sections.* Cambridge, Barnes and Noble (reprint), 1961.

[31] HEATH, SIR THOMAS L., *The Works of Archimedes.* New York, Dover Publications (reprint), 1953.

[32] HEATH, SIR THOMAS L., *A History of Greek Mathematics*, Vols. I&II. Oxford, London, etc., Oxford University Press, 1965.

[33] HOFMANN, J.E., *Frans van Schooten der Jüngere.* Wiesbaden, Franz Steiner Verlag, 1962.

[34] HOGENDIJK, J.P., Kegelsneden in de Griekse Oudheid. *CWI syllabus* **40**, pp. 1–20. Amsterdam, Centrum Wiskunde & Informatica, 1995.

[35] HOGENDIJK, J.P., Book I of Jan de Witt's Elementa Curvarum Linearum and the Conics of Apollonius. *Nieuw Archief voor Wiskunde* (4), XVII, (1999), pp. 453–463.

[36] HULTSCH, F., *Pappi Alexandrini Collectionis quae supersunt*, Vol. I&II. Berlin, Weidmann, 1877.

[37] HUYGENS, CHRISTIAAN, *Œuvres complètes de Christiaan Huygens*, publiées par la Société Hollandaise des Sciences, Vols. I–XXII. Den Haag, 1888–1950.

[38] JAPIKSE, N., *Johan de Witt.* Amsterdam, Meulenhof, 1915.

[39] KATZ, V.J., *A History of Mathematics, an Introduction*. New York, Harper Collins College Publications, 1992.

[40] KLEIN, J., *Greek Mathematical Thought and the Origin of Algebra*. Cambridge Massachusetts, and London, The M. I. T. Press, 1968.

[41] KLINE, M., *Mathematical Thought from Ancient to Modern Times*. New York, Oxford University Press, 1972.

[42] KNORR, W.R., *The Evolution of the Euclidean Elements*. Dordrecht, Reidel, 1975.

[43] KNORR, W.R., *The Ancient Tradition of Geometric Problems*. New York, Dover Publications Inc., 1993.

[44] LEIBNIZ, G.W., Nova methodus pro maximis et minimis (New method for maximums and minimums). In *Acta Eruditorum*, 1684; translated in Struik, D.J., 1969. *A Source Book in Mathematics, 1200–1800*. Harvard University Press: 271–81.

[45] MAANEN, J.A. van, *Facets of Seventeenth Century Mathematics in the Netherlands*. Utrecht, Elinkwijk, 1987.

[46] MAHONEY, M.S., *The Mathematical Career of Pierre de Fermat (1601–1665)*. Princeton, New Yersey, Princeton University Press, 1973.

[47] MORROW, G., *Proclus; A Commentary on the First Book of Euclid's Elements*. Princeton, New Jersey, Princeton University Press, 1970.

[48] RABUEL, CLAUDE, *Commentaire sur la Géométrie de M. Descartes*. Lyons, 1730.

[49] ROWEN, H.H., *John de Witt, Grand Pensionary of Holland, 1625–1672*. Princeton, New Jersey, Princeton University Press, 1978.

[50] ROWEN, H.H.., *John de Witt, Statesman of the True Freedom*. Cambridge, Cambridge University Press, 1986.

[51] SCHOOTEN JR., F. VAN, *Geometria à Renato Des Cartes Anno 1637 Gallicè edita; postea autem unà cum Notis Florimondi de Beaune, (...) in Latinam linguam versa, & Commentariis illustrata, opera atque studio Francisci à Schooten*. Leiden, Jan Maire, 1649. Revised and extended edition in 1659.

[52] SCOTT, J.F., *The Mathematical Work of John Wallis, D.D., F.R.S. (1616–1670)*. London, Taylor and Francis, 1938.

[53] SCOTT, J.F., *The Scientific Work of René Descartes*. London, Taylor and Francis, 1976.

[54] STECK, M., *Proklos Diadochus (410–485); Kommentar zum Ersten Buch von Euklids "Elementen."* Halle, Deutsche Akademie der Naturforscher, 1945.

[55] STEINER, JAKOB, *Geometrical Constructions with a Ruler* (a translation of his 1833 book). New York, Scripta Mathematica, 1950.

[56] THAER, C., *Euklid, die Elemente. Buch I–XIII*. Herausgegeben und ins Deutsche Übersetzt von Clemens Thaer. Darmstadt, Wissenschaftliche Buchgesellschaft, 1962.

[57] THOMAS, IVOR, *Greek Mathematics*, Vols. I&II, Cambridge Mass., Harvard University Press, 1957.

[58] TOOMER, G.J., *Diocles on Burning Mirrors*. New York etc., Springer-Verlag, 1976.

[59] VEREECKE, PAUL, *Les Coniques d'Apollonius de Perge. Œuvres traduites pour la première fois du Grec en Français*. Bruges, Desclée de Brouwer et Cie, 1923.

[60] VEREECKE, PAUL, *Pappus d'Alexandrie, la Collection Mathematique*. Tome I, Tome II. Bruges, Desclée de Brouwer et Cie, 1933.

[61] VIDELA, CARLOS, R, On Points Constructible from Conics. *Mathematical Intelligencer* **10** (1997), pp. 53−57.

[62] VIÈTE, FRANÇOIS, Opera Mathematica (reprint of ed. 1646). Hildesheim, Georg Olms Verlagsbuchhandlung, 1970.

[63] VRIES HK. DE, Historische Studiën II, *Nieuw Tijdschrift voor Wiskunde*. 12e Jrg. 1924/25, pp. 260–288

[64] WAERDEN, B.L. VAN DER, *Science Awakening*. Groningen, Noordhoff, 1954.

[65] WAERDEN, B.L. VAN DER, *A History of Algebra*. New York etc., Springer-Verlag, 1985.

[66] WALLIS, JOHN, *Opera Mathematica*, Vols. I–III. Oxford, Th. Robinson, 1693–1699.

[67] WALLIS, JOHN, *Opera Mathematica*, Vols. I–III. Reprint introduced by C.J. Scriba. Hildesheim-New York, Georg Olms Verlagsbuchhandlung, 1972.

[68] WITT, JAN DE, *Elementa Curvarum Linearum −Liber Primus*, Tekst, vertaling inleiding en commentaar door A.W. Grootendorst. Amsterdam, Centrum Wiskunde & Informatica, 1997.

[69] WITT, JAN DE, *Elementa Curvarum Linearum, Liber Primus*, Text, translation, introduction and commentary by Albert W. Grootendorst (with the help of Miente Bakker). Sources and Studies in the History of Mathematics and Physical Sciences, New York etc., Springer-Verlag, 2000.

[70] WICKEFOORT CROMMELIN, H.S.M. VAN, *Johan de Witt en zijn Tijd*.
Amsterdam, Nederlandse Algemene Maatschappij van Levensverzekering "Conservatrix," 1913.

[71] ZEUTHEN, H.G., *Geschichte der Mathematik im Altertum und Mittelalter*. Kopenhagen, A.F. Host, 1896.

[72] ZEUTHEN, H.G., *Die Lehre von den Kegelschnitten im Altertum*. Hildesheim, Georg Olms Verlagsbuchhandlung, 1966

Sources of illustrations

Pictures on pages vi and x

All these pictures are freely available on the internet.

Jan de Witt   Painting by Jan de Baen (1633–1702)
Source: Rijksmuseum Amsterdam
http://commons.wikimedia.org/wiki/File:Grand_Pensionary_Johan_
de_Witt.jpg

René
Descartes
Painting by Frans Hals (ca 1583–1666)
Source: Louvre, Museum, Paris
http://en.wikipedia.org/wiki/File:Frans_Hals_-
_Portret_van_René_Descartes.jpg

Pierre de   http://commons.wikimedia.org/wiki/File:Pierre_de_Fermat.jpg
Fermat

Frans van   http://commons.wikimedia.org/wiki/File:Frans_van_schooten_jr.jpg
Schooten, jr.

François   http://commons.wikimedia.org/wiki/File:Francois_Viete.jpg
Viète

Other pictures in Chapters 1–5

Most figures were drawn by Tobias Baanders (CWI, Amsterdam), with the
following exceptions
- Figure 5.8 was inspired by Heath [30]
- Figure 5.9 was derived from Descartes [16]
- Figure 5.13 (drawn by Jan Aarts) was derived from J.D. LAWRENCE, *A catalog of Special Curves*, New York, Dover publications, 1972

Index